ADVANCES IN
BIOSENSORS

Volume 5 • 2003

PERSPECTIVES
IN BIOSENSORS

Previous volumes published by:
JAI Press INC.
100 Prospect Street, Stamford, Connecticut
USA

ADVANCES IN BIOSENSORS

PERSPECTIVES IN BIOSENSORS

Editors: BANSI D. MALHOTRA
Biomolecular Electronics &
Conducting Polymer Research Group
National Physical Laboratory
New Delhi, India

ANTHONY P.F. TURNER
Cranfield Biotechnology Centre
Cranfield University
Bedfordshire, England

VOLUME 5 • 2003

JAI

An Imprint of Elsevier Science
Amsterdam – Boston – London – New York – Oxford – Paris
San Diego – San Francisco – Singapore – Sydney – Tokyo

ELSEVIER SCIENCE B.V.
Sara Burgerhartstraat 25
P.O. Box 211, 1000 AE Amsterdam, The Netherlands

First edition 2003

Second impression 2004

ISBN: 0-444-51337-X
ISSN: 1061-8945

Transferred to digital printing 2006

CONTENTS

LIST OF CONTRIBUTORS

N.K. Chaudhury	Institute of Nuclear Medicine and Allied Sciences Brig. S.K. Majumder Rd. Delhi 110 054 India
Pratima Chauhan	Biomolecular Electronics and Conducting Polymer Research Group National Physical Laboratory Dr. K.S. Krishnan Road New Delhi 110012 India
M.N. Kamalasanan	Biomolecular Electronics and Conducting Polymer Research Group National Physical Laboratory Dr. K.S. Krishnan Road New Delhi 110012 India
N.G. Karanth	Fermentation Technology and Bioengineering Department Central Food Technological Research Institute Mysore 570013 India
Arun Kumar	Biomolecular Electronics and Conducting Polymer Research Group National Physical Laboratory Dr. K.S. Krishnan Road New Delhi 110012 India

Bansi D. Malhotra Biomolecular Electronics and Conducting Polymer Research Group
National Physical Laboratory
Dr. K.S. Krishnan Road
New Delhi 110012
India

Chanchal K. Mitra School of Life Sciences
University of Hyderabad
Hyderabad 500 046
India

Mangu Singh Biomolecular Electronics and Conducting Polymer Research Group
National Physical Laboratory
Dr. K.S. Krishnan Road
New Delhi 110012
India

Vibha Saxena Biomolecular Electronics and Conducting Polymer Research Group
National Physical Laboratory
Dr. K.S. Krishnan Road
New Delhi 110012
India

P. Shantilatha School of Life Sciences
University of Hyderabad
Hyderabad 500 046
India

C. Raman Suri Institute of Microbial Technology
Sector 39-A
Chandigarh 160 036
India

M.S. Thakur Fermentation Technology and Bioengineering Department
Central Food Technological Research Institute
Mysore 570013
India

Shailly Varma School of Life Sciences
University of Hyderabad
Hyderabad 500 046
India

PREFACE

There is a worldwide effort towards the development of bioanalytical devices that can be used for detection, quantification and monitoring of specific chemical species. In this context, biosensors represent an emerging trend in the diagnostics industry. A biosensor is a device that has a biological sensing element either intimately connected to or integrated within a transducer. The aim is to produce a digital electronic signal that is proportional to the concentration of a specific chemical or a set of chemicals. Biosensors are specific, rapid, cost-effective and easy to use devices that can be employed with minimal sample treatment. Biosensors have applications in many areas such as biotechnology, healthcare, pollution monitoring, food and agriculture product monitoring, the pharmaceuticals industry and defense.

The development of biosensors is still an open field and much remains to be done before many of these bioelectronic devices become commercialized. Keeping this in view, it was felt that this reference text devoted to the principles and applications of biosensors would fulfill a niche demand of academic institutes, research laboratories and the rapidly developing biosensor industry.

Advances in Biosensors: Perspectives in Biosensors has been designed to discuss novel ways that can be used to fabricate biosensors for a variety of applications. P. Shantilatha *et al.* discuss the fundamental principles used in the design of a biosensor (**Chapter I**). N.K. Chaudhuri presents a brief review of research and development of optical biosensors (**Chapter II**). Vibha Saxena *et al.* have shown that it is possible to fabricate electrochemical biosensors that can detect and manipulate single biomolecules **(Chapter III)**. Interestingly, the electrochemical biosensors described have been successfully employed in the clinical diagnostics industry. The biological application of sol–gel chemistry is quite a new field. A. Kumar *et al.* show that sol–gel chemistry provides a versatile and simple route for the development of electrochemical and optical biosensors for many types of clinical analysis **(Chapter IV)**. Thakur and Karanth show that biosensors have a tremendous scope for monitoring food and beverages as well as in food and water-borne pathogens analysis, and should be vigorously pursued **(Chapter V)**. C. Raman Suri discusses immunosensors based on antibody/antigen binding that can be successfully utilized for the monitoring of widely used pesticides for increasing agricultural production **(Chapter VI)**.

We are grateful to all the members of both the Biomolecular Electronics Group of the National Physical Laboratory, New Delhi, India and the Biotechnology Centre, Cranfield University, Silsoe, U.K., for many suggestions and discussions during the preparation of the book.

Advances in Biosensors: Perspectives in Biosensors has been realized with invaluable contributions from experts who have been active in their respective fields of research for decades. We are thankful to all of them for their active participation in this important project. Special thanks are to Mrs Linda Ball (Cranfield University at Silsoe), Mr Jeroen Soutberg and Mr Derek Coleman and the staff at Elsevier who have worked hard to see that everything in the book is correct and is completed on time.

Bansi D. Malhotra
Scientist-In-Charge
Biomolecular Electronics & Conducting Polymer Research Group,
National Physical Laboratory, New Delhi-110012, India
E-mail: bansi@csnpl.ren.nic.in

A.P.F. Turner
Head, Cranfield University at Silsoe
Cranfield, England, MK45 4DT, U.K.
E-mail: a.p.turner@cranfield.ac.uk

Chapter 1

DESIGNING A SIMPLE BIOSENSOR

P. Shantilatha, Shailly Varma and Chanchal K. Mitra

OUTLINE

Advances in Biosensors
Volume 5, pages 1–36.

1. INTRODUCTION

There is a need for simple, rapid and reagentless method for specific determination, both qualitative and quantitative, of various compounds in various applications. A large number of such applications exist in clinical diagnostics and in biotechnology. The requirement for accurate and fast chemical intelligence is particularly conspicuous in human health care. It has also become increasingly important in several other areas like veterinary medicine, agrifoods, horticulture, pharmaceuticals, petrochemical industries, environment surveillance, defense and security. Conventional chemical determinations suffer from lack of specificity and a host of interferences, apart from being slow and expensive.

1.1. What is a Biosensor?

A biosensor represents the convergence of two contrasting disciplines – *it combines the specificity and selectivity of biological systems with the computing power of the microprocessor*. A biosensor is a device incorporating a biological recognition element either intimately connected to or integrated with a transducer. Classically biosensors can be defined as analytical devices in which a biological component is coupled with a transducer to convert a biological signal into an electrical one. The field of biosensor technology holds many exciting analytical (both qualitative and quantitative) applications for enzymes (or other biomolecules). It is a very diverse field, making use of expertise in biochemistry, immunology, optical physics, electrochemistry, materials science, semiconductors, electronics and a myriad of other science and engineering disciplines. A schematic representation of a typical biosensor can be seen in Figure 1.

The concept of biosensor was proposed more than four decades ago. Today there are more than 60 commercial devices available for about 120 different analytes. Among these are sensors for low-molecular-mass substances, sensors for enzyme components and sensors for macromolecules (such as viruses, micro-organisms) (Meixner 1995). Around 12 to 15 million US$ per year is being spent on analytical purposes worldwide. They include usage of enzymes in clinical chemistry, food and cosmetic industry and biotechnology for

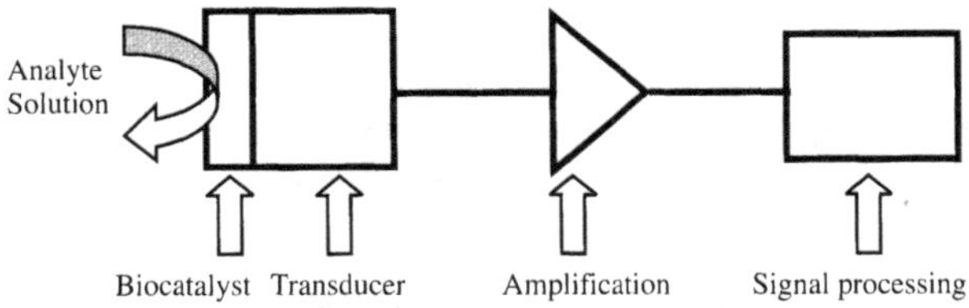

Figure 1. Schematic diagram of a biosensor showing the various components.

the measurement of 80 different substances. Fundamental problems pertaining to cost of biomolecules and intermediates, storage, operational stability etc. may contribute to such relatively slow development. Immobilization of the biomolecule permits reuse of costly material and allows significant simplification of analytical apparatus. The various components are as follows:

(1) *Biologically sensitive element or biocatalyst*: this is generally an immobilized biological system or component that is able to specifically recognize the target molecule among many others. The biocomponents mainly used in the design of biosensors are enzymes, antibodies, organelles, bacteria, whole cells, tissue slices, etc.
(2) *Transducer*: the transducer converts the biorecognition into a measurable electrical signal such as current or voltage values. The output from the transducer can be displayed directly or can be further processed by a microprocessor or amplifier or other signal processing techniques. Some of the commonly used transducers have been listed in Table 1 (below).
(3) *Amplifier and signal processing*: these are standard electronic components used to process the transducer signal into a display, which is mostly a concentration in some convenient units. Sometimes the display can be simple, indicating a go/no-go test.

The early biosensors used conventional electrodes (pH electrode; oxygen electrode; ion-selective electrodes, etc.) as a transducer in conjunction with a suitable biocatalyst. The enzyme electrode is a combination of a suitable electrode sensor with an immobilized (insolubilized or trapped) enzyme, which provides a highly selective and sensitive method for the determination of a given substrate. One of the early biosensors for glucose uses the oxygen electrode for the measurement of the decrease in the oxygen concentration.

The early biosensor: Clark and Lyons (1962) introduced the concept of the 'soluble' enzyme electrode. The next development was reported

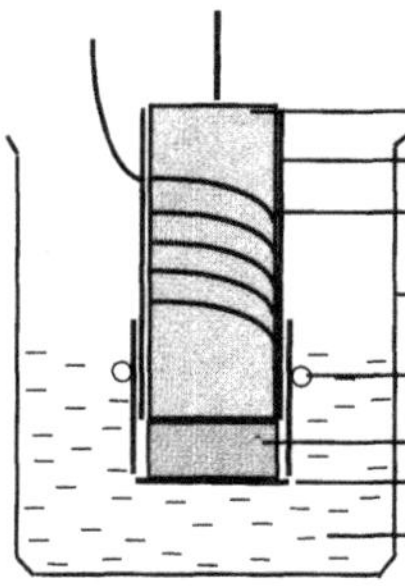

Figure 2. Schematic of the original glucose biosensor. The Ag/AgCl counter electrode is wrapped around the body of the working electrode. The base electrode is set at a potential of −0.6 V (suitable for oxygen reduction) and the current is measured amperometrically.

by Updike and Hicks (1967) who immobilized glucose oxidase (GOD) in a gel over a polarographic oxygen electrode in order to measure the concentrations of glucose in biological fluids. Early enzyme electrodes used transducers that were voltammetric or amperometric probes, i.e., either the current produced upon application of a constant applied voltage (amperometric method) or spontaneous production of a potential under zero current condition (voltammetric or potentiometric method) was measured. For example, a pH electrode is a voltammetric probe, whereas an oxygen electrode is an amperometric probe. The first potentiometric (with no external potential applied, the voltage produced is monitored) enzyme electrode was described by Guilbault and Montalvo for urea in 1969. Since then, over 100 different electrode designs have appeared in the literature (Guilbault 1984). The early glucose biosensor is shown schematically in Figure 2. The diagram shows all the essential components. The transducer is an oxygen electrode, i.e., the decrease in oxygen tension due to the oxidation of glucose by the enzyme is actually measured. This sensor suffers from several drawbacks: (i) if other redox species are present (that can be reversibly reduced at the operating potential) then some interference is expected and (ii) if the initial oxygen concentration is low for some reason, the readings may not be accurate. This is an example of a first-generation biosensor and modern biosensors for glucose determination are far more resistant to interference and do not critically depend on the initial oxygen concentration. The response of the electrode is relatively slow, as a considerable barrier to diffusion exists. In Figure 2, the 'biocatalyst' can be an enzyme paste, a tissue slice, an enzyme immobilized on a matrix, etc. The insulating sleeve on the side of the electrode ensures that the solution contacts the electrode only through the biocatalyst. The base electrode can be either a Pt or Au or a glassy carbon rod. This base electrode acts as an oxygen electrode and can be used in the amperometric mode at a suitable bias potential. A semipermeable membrane, e.g., a thin nylon, cellophane or other suitable film, holds the biocatalyst tightly pressed in position against the base electrode. The 'O' ring can be a rubber band to keep the membrane stretched tightly over the catalyst. The auxiliary electrode (Ag/AgCl) is wound around the main electrode. The current passing through the electrode is proportional to the oxygen concentration at the electrode surface. This set-up can however be tailored according to the specificity required. Using suitable enzymes (that catalyze a reaction in which oxygen is consumed), this can be used for a number of different substrates.

Membranes introduce a diffusional barrier that slows down the response. However, they are highly desirable as they prevent the deposition of the biological macromolecules (e.g., proteins and polysaccharides) on the electrode surface. With glucose oxidase enzyme mixed with agar, gelatin, BSA or some other indifferent matrix to form a paste or a gel, a simple immobilized biocatalyst can be made. The paste or gel can be applied as a thin layer on top of the electrode and protected by a semipermeable membrane. Each electrode needs frequent calibration and often there is considerable drift in the current values. This analyzer can measure the glucose content of whole blood, plasma or serum, requiring very little amounts of sample.

In enzyme electrodes, the enzyme is immobilized close to the electrode surface. This reduces the amount of material required to perform a routine analysis and eliminates the need for frequent assay of the enzyme preparation in order to obtain reproducible results. Furthermore, the stability of the enzyme is often improved when it is incorporated in a suitable gel matrix. For example, an electrode for the determination of glucose prepared by covering a platinum electrode with chemically bound glucose oxidase could be used for more than 300 days (Guilbault and Lubrano 1973). The enzyme electrode operates via a five-step process: the substrate must diffuse through the semipermeable membrane to the biocatalyst; the substrate must be transported to the active site of the biocatalyst; biochemical reaction occurs at the active site and products are formed; the active product diffuses across the biocatalyst to the surface of the transducer; product concentration is measured at the electrode surface electrochemically. All of the above processes can be rate-limiting, and the slowest step determines the overall rate of the whole system (i.e., the current produced). Conditions should be optimized so that the current produced is proportional to the substrate (glucose) concentration.

1.2. Classification of Biosensors Based on Type of Transduction

There are several types of transducers. Some of the commonly employed methods of transduction are listed in Table 1. The major transduction methods are briefly described here.

Table 1. Types of transducers commonly used in biosensors

Transducer	Examples
Electrochemical	
Amperometric	Clark oxygen electrode, mediated electrode systems
Potentiometric	Redox electrodes, ion selective electrodes, field effect transistors, light addressable potentiometric sensors
Conductometric	Platinum or gold electrodes for the measurement of change in conductivity of the solution due to the generation of ions.
Optical	Photodiodes, waveguide systems, integrate optical sensors.
Acoustic	Piezoelectric crystals; surface acoustic devices.
Calorimetric	Thermistor or thermopile

1.2.1. Electrochemical transducers:

These are based on electrochemically monitoring the response of the biosensors. There are three types: potentiometric, amperometric and conductometric.

(i) Potentiometric transduction: A potentiometric biosensor monitors the potential under zero current conditions. The potential generated is directly proportional to the logarithm of the analyte concentration. The basis of this type of electrochemical monitoring is the

Nernst equation, which relates the electrode potential (E) to the concentration of the oxidized and reduced species:

$$E = E^{0\prime} + \frac{RT}{nF}\ln\left(\frac{[\mathrm{Ox}]}{[\mathrm{Red}]}\right).$$

(ii) Amperometric transduction: Amperometry encompasses a group of electroanalytical techniques. It monitors the current generated at a fixed bias potential. Linear relation between the concentration of the analyte and the current generated is obtained. Several techniques like cyclic voltammetery, flow injection analysis studies, etc. are commonly employed. Many biosensors are based on amperometric-type detection.

(iii) Conductometric transduction: This is a technique where the changes in ionic concentrations are measured. If the biocatalyst produces ionic products, or consumes ions, and the support solution has low electrical conductivity, this is often a convenient and simple technique.

1.2.2. Optical Transducers

Optical methods have been used classically to monitor analyte concentrations. Properties like absorption, refractive indices, fluorescence, phosphorescence, chemiluminescence, etc., can be used in order to monitor the biological recognition in biosensors. The devices can be miniaturized by using optical fibres, which act as light guides. The detectors are often semiconductor photodiodes. These devices are often useful for remote analysis as the light signal is resistant to electrical noise.

1.2.3. Piezoelectric Transducers

Piezoelectricity is a property possessed by anisotropic (lacking symmetry) crystals such as quartz. When such a crystal is stressed, it causes an electric field to develop. In a similar fashion, when an electric field is applied to the crystal, it undergoes a mechanical deformation. Under an alternating voltage, the crystal oscillates with the applied frequency. The natural frequency of the crystal is related to the mass and the elastic constants of the crystal. Phenomena such as electrodeposition or dissolution, where a change in mass of the crystal occurs, can be studied. Many biosensors based on microbalances are also used.

1.2.4. Thermistors

They are devices used to monitor the enthalpy change of an enzyme-based reaction. If the enthalpy change in the biocatalytic process is significant, then the temperature of the transducer (thermistor) is changed and this change can be monitored. A thermistor is nothing but a miniature resistance thermometer with high sensitivity.

1.2.5. Semiconductor-based Electrodes

These are transistor-like devices (usually npn type), and the most common configuration is the field effect transistor (FET). The biological element can be immobilized at the surface of the gate of the FET to obtain EnFET (Enzyme Field Effect Transistor). They operate in the potentiometric mode.

1.3. Classification of Biosensors Based on Biological Element

1.3.1. Enzyme Sensors

As mentioned earlier the use of glucose-based sensors is studied extensively. Usually the conventional method of detecting oxygen used or the peroxide molecule released is monitored. Apart from this there are sensors relying on mediator-based systems (*vide infra*). The use of reconstituted enzyme paste with a direct molecular wire to the electrode surface results in a better electron transfer between the enzyme and the electrode, as seen from our results.

In addition to glucose oxidase, work on several other enzyme sensors is in progress. An important example is the penicillin electrode used to monitor the penicillin content of fermentation broths. This electrode is based on a pH probe coated with immobilized penicillinase (Kobos 1980). The probe responds to the changes in the $[H^+]$ arising from the dissociation of penicilloic acid formed.

Another important metabolite often monitored is urea levels in blood (blood urea and nitrogen, BUN) for assessing kidney damage or failure. This sensor employs urease enzyme and usually relies on the pH changes

Table 2. Examples of biosensors

Analyte	Biocatalyst	Product/ transducer	Immobilization	Response time
Alcohol	Alcohol dehydrogenase	O_2	Cross-linked	1–2 min
Arginine	*Streptococcus faecium*	NH_3 (pH)	Physically entrapped	20 min
Cholesterol	*Nocardia erthyropolis*	O_2	Cross-linked	35–70 s
D-Glucose	Glucose oxidase	O_2	Dialysis membrane	1 min
Glutamate	Glutamatedecarboxylase	CO_2 (pH)	Physically entrapped	10 min
NAD^+	NADase and *E. coli*	NH_3 (pH)	Polyacrylamide gel	5–10 min
Nitrate	*Azotobacter vinelandii*	NH_3 (pH)		7–8 min
Penicillin	Penicillinase	H^+ (pH)	Polyacrylamide gel	15–30 s
Urea	Urease	NH_4^+ (pH)		20–40 s

[a] Adapted from Wilson and Walker (1996).

during urea hydrolysis. Petersson *et al.* (1987) have developed a urea analyzer for undiluted samples by using an urease-covered ammonium-ion selective electrode in a flow injection analysis system.

Lactate determination is important for the diagnosis of shock and myocardial infracture, neonatology and in sports medicine. An example of a lactate sensor is polyurethane immobilized lactate dehydrogenase used for whole-blood lactate determination by glukometer and ECA20 (ESAT660) (Schubert *et al.* 1989). Some of these biosensors are listed in Table 2.

1.3.2. Immunosensors

Electrochemical immunoassay is also gaining a lot of importance. Excellent detection limits have been achieved with small sample volume. Modern electrochemical techniques have stimulated the development of electrochemical immunosensors. Enzymes can be used as an effective label in immunoassay methods. Homogeneous enzyme immunoassays rely on the change in the intensity of the label signal that occurs when Ag^* (labeled) binds with Ab to form $Ab{:}Ag^*$. The ability to distinguish electrochemically Ag^* from $Ag^*{:}Ab$ is used without the need for separation.

Broyles and Rechnitz (1986) have designed an immunoassay for drug antibodies. The assay is based on inhibition of enzyme–antigen conjugate (E-Ag) to the substrate by corresponding Ab, which is the analyte. Activity is monitored by amperometric detection of the rate of NADH oxidation at the platinum electrode. A magnetic working electrode was used by Robinson *et al.* (1985) in order to separate the bound and free enzyme label to monitor the electrochemical response. Monoclonal antibodies for human chorionogonadotropin (hCG) was the model system. Magnetic particles were used as the solid phase. When analyte is present enzyme label is bound to the electrode via a series of species (including hCG Ab) connected to the magnetic particle. Separation of the bound and free label results from the localization of the particles at magnetic electrodes. GOD activity was measured at the electrode electrochemically using dimethylaminomethylformamide as the electron mediator.

There have been reports on direct attachment of the Ab on the carbon electrode surface so that the magnetic electrode is not necessary (Robinson *et al.* 1986). A biosensor was developed for online measurement of progesterone in bovine milk and detection of estrus by Claycomb and Delwiche (1998). It is based on a modified EIA (Enzyme Immunoassay) format for molecular recognition, designed to operate online in a dairy parlor, with a response time of approximately 8 minutes.

1.3.3. Cell-based Sensors

Significant work has been done on electrochemical sensors, amperometeric as well as potentiometric, using microorganisms (Corcoran and Rechnitz 1985). The cells are fixed at the electrode surface between the sensing membrane of the probe and the retaining membrane. A semipermeable membrane is used for this purpose. The general principle of operation and response of cell-based sensors is similar and comparable to that of enzyme electrodes prepared with electrochemical detectors. Nonetheless, these devices have some advantages over the enzyme electrodes (e.g., reduced cost, higher catalytic activity and improved stability). They can be regenerated in growth medium to replenish the supply of active cells therefore the lifetime of the electrode is increased. The major disadvantage is the poor selectivity of microbial sensors due to the multireceptor behavior of the intact cells. They also require a longer recovery time between measurements due to increased number of cells on the electrode surface resulting in diffusion constraints. This point is not of concern in single-use disposable sensors. This kind of sensors thus can be used in fermentation processes, clinical analysis of biological fluids.

Excellent reviews on biosensors can be found in articles written by Thompson and Krull (1991) and Frew and Hill (1987). The design of biosensor-based analyzer mainly depends on its specific use in medicine, industry, environment or defense. Within each sector, the requirement may also be totally different due to the specific site where the system has to be used. The application of biosensor for clinical analysis has been extended to bedside monitoring or to direct implantation into an organ, thanks to biomedical engineering. Yet, serious difficulties have to be overcome, namely for biocompatibility, lifetime, sterilization, etc.

Overall, for biosensors the following lines of applications are emerging:

- Medical diagnostics, including patient monitoring in intensive medicine and bed-side analysis.
- Environmental analysis, above all waste-water and water pollution control.
- Food quality control, for example, for purity and freshness.
- Process control and substance testing, and other utilization in the chemical–pharmaceutical industry.
- Drug detection.

1.4. Three Generations of Biosensors

Over the years, biosensors have come a long way. In the case of amperometric enzyme electrodes an oxidoreductase is typically involved and glucose oxidase is one of the most widely and extensively studied enzymes for use in glucose sensors. We have the following generations of biosensors as exemplified by the development of the glucose sensor.

Case 1 (first generation). The enzyme glucose oxidase (GOD) acts on glucose to produce H_2O_2 in presence of oxygen. Thereby, the oxygen concentration at the electrode surface is decreased. The redox group in GOD is FAD (Flavin adenine dinucleotide) that undergoes a $2e^-$ $2H^+$ reduction. The overall reaction is described as follows:

$$\mathrm{GOD_{OX}} + \text{glucose} \rightarrow \mathrm{GOD_{RED}} + \text{gluconolactone},$$

$$\mathrm{GOD_{RED}} + O_2 \rightarrow \mathrm{GOD_{OX}} + H_2O_2.$$

Both FAD and $FADH_2$ are highly soluble in water but are strongly bound to the enzyme and do not exchange with FAD/$FADH_2$ in solution. In the above reaction we have therefore shown the redox process of the FAD as that of the enzyme. The decrease in O_2 concentration is proportional to the glucose concentration (under suitable experimental condition). Alternatively, the product formed (H_2O_2) can also be measured at a suitable potential.

Case 2 (second generation). One potential technical difficulty in using an oxidoreductase is oxygen limitations under conditions of low oxygen tension. This can be particularly acute for *in vivo* applications where, on a molecular basis, the concentration of unbound oxygen may be lower than that of the analyte, resulting in stoichiometric limitations of the enzyme reaction by oxygen. One method to overcome this problem is to replace the oxygen with an artificial mediator. Electron mediators enhance the rate at which the transfer of electrons takes place (Carr and Browers 1980, Cass *et al.* 1984). Their role is to shuttle electrons between the enzyme and electrode as shown in Figure 3. However, one of the major problems is that it is generally difficult to tailor the redox potential of the molecule and to have high efficiencies (so that the mediator molecule can do the same job repeatedly, as a catalyst). If the potentials do not match well, then there would be energy inefficiency and the mediator may undergo side reactions. Therefore, the selected mediator molecule should have an appropriate redox potential under the specified experimental conditions.

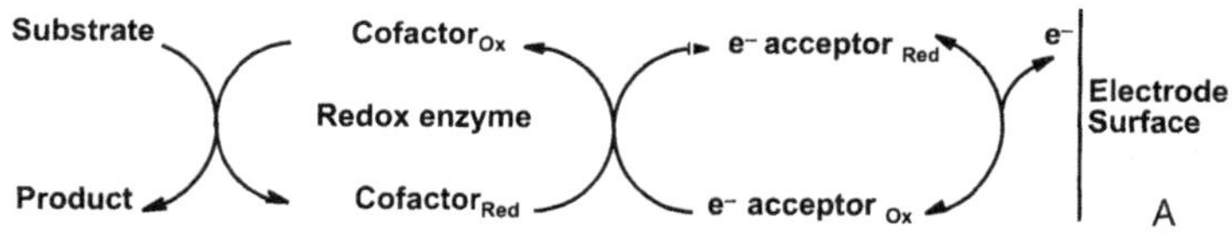

Figure 3. Electron transport shown schematically for a mediated reaction.

The selection of a suitable mediation is often a trial and error process. The following reaction scheme can be proposed with a mediator:

$$GOD_{OX} + \text{glucose} \rightarrow GOD_{RED} + \text{gluconolactone},$$

$$GOD_{RED} + Med_{OX} \rightarrow GOD_{OX} + Med_{RED},$$

$$Med_{RED} \rightarrow Med_{OX} + ne^{-} \text{ (at the electrode surface).}$$

Some of the electron mediators commonly employed are ferro/ferri cyanide, hydroquinone, ferrocene, various redox dyes, etc. One of the limitations here is that any other electroactive redox species that may be present can cause some interference. Oxygen can also compete with the mediators for electrons but the kinetics (applied potential) and the chemical nature of the dye will determine the degree of interference.

Case 3 (third generation). In a mediatorless case, the electrode exchanges electrons directly with the enzyme and hence with the substrate. Thus, the redox potential of the enzyme is realized. However, this is often difficult as the active site of the enzyme is physically distant from the electrode surface. On the other hand, if the enzyme is very close to the electrode surface, it will have too little conformational freedom and poor access to the substrate, and the environment will be dehydrated. Therefore, it is only by very careful and judicious choice that one can get direct electron transfer to and from the electrode. Another choice, recently advocated, is the use of molecular wires that can be used to connect the enzyme active site to the electrode surface using a chemical link. This reaction scheme can be described as follows:

$$\mathrm{GOD_{OX}} + \text{glucose} \rightarrow \mathrm{GOD_{RED}} + \text{gluconolactone},$$

$$\mathrm{GOD_{RED}} \rightarrow \mathrm{GOD_{OX}} + ne^- \ \text{(to the electrode)}.$$

This process does not involve the mediator or oxygen and depends on the direct oxidation–reduction of the enzyme by the electrode. This is the preferred way, but at the same time the most complex. The elimination of interference, one of the major objectives, is one of the greatest advantages in this design.

A steady-state electrochemical signal is reached after a short delay, when the rate of product formation equals the rate at which the product diffuses out of the enzyme membrane. Alternatively, a kinetic measurement can be used in which the rate of change of the electrochemical signal is monitored as a function of time. Therefore the measurement of the signal at the correct time is an important consideration.

Third-generation biosensors are promising in order to provide a powerful and inexpensive alternative to conventional analytical strategies for assaying chemical species present in complex matrices. This is because of their ability to discriminate the target analyte from a host of inert and potentially interfering species without the requirement for separating and subsequently identifying all the constituents of the sample. This can be simply achieved by using several different enzymes on the same transducer and scanning the potential (as in polarography) so as to detect different components provided the potentials are sufficiently distinct.

1.5. Chemical Sensors and other Sensors

Chemical sensors are a closely related group of devices that are used to detect the compounds of interest. These devices use small organic molecules instead of biological elements to detect the analyte (mainly gases).

1.5.1. Electronic Noses

A good example of a chemical sensor is the electronic nose. The concept of electronic nose was first introduced by Persaud and Dodd (1982). Since then, a number of artificial nose devices have been developed. The artificial olfactory system devised by White *et al.* (2002), the Tufts medical school nose (TMSN), finds application in the detection of land mines. It can detect nitro compounds like dinitrotoluene, trinitrotoluene and methanol, amongst others. An electronic nose incorporates two main features: (a) an array of sensors with rather broad specificity along with (b) a pattern recognition method for processing the multitude of sensor outputs.

1.5.2. Sensors for Heavy Metals

Porphyrin derivatives are a promising class of chelating agents being well established for the spectroscopic determination of heavy metal ions. They can be coupled with sensitive optical detecting techniques. Porphyrins have been stably immobilized on thin films of Nafion (strongly acidic cationic exchanger) for detecting cadmium and mercury. With a measuring time of 10 min, a detection limit of 5 μg/L for Cd (II) and 30 μg/L for Hg (II) was observed.

1.5.3. Sensors for the Detection of Toxic Gases

Metalloporphyrins can also be used to study the spectral change using optoelectrochemical gas detection techniques. Sensors developed for the determination of NH_3 and SO_2 are based on metalloporphyrins, which are immobilized on thin films (<1 μm) of porous glass prepared by a sol–gel technique.

1.6. Instrumentation

Biosensors are the detector unit of a large and complex system. The hand-held unit commonly used by diabetics for the measurement of glucose concentration is certainly inconvenient in a hospital set-up. In a hospital (clinical laboratory) set-up, one would like to get as much as information possible from the same sample in a short period. In addition, the instrument must be designed so as to allow rapid sequential measurement of a large number of samples. Therefore, multisensor arrays are often used in a large set-up.

1.6.1. Stationary Portable Systems

About 4% of the population in industrialized countries suffers from diabetes, and rapid and reliable methods for the detection of blood glucose are of utmost importance for diabetes screening and treatment. It has been anticipated that more than 500 million glucose sensors are made world wide.

One glucose sensor is manufactured by Genetics International (UK) (Bartlett *et al.* 1987). The sensor is based on ferrocene-modified GOD electrode strips. For the determination of glucose a drop of blood is applied on to the strip, which is then introduced into a pen-sized readout instrument. The precision in the normal range is 3.9%. Lower precision is found in hypoglycemic range and an appropriate correction has been proposed. The simple handling of the instrument makes it well suited for use in doctor's clinics and by patients.

1.6.2. Flow Systems Suitable for Large Number of Samples: Sequential Semi-automated Analysis

The determination of chemical parameters is of great interest in all kinds of bioprocesses. In the bioprocess industry, on-line measurements of nutrients, metabolites and products to be used for process control are most important. For the solution of such problems, online analytical methods are in demand and the use of either FIA (flow injection analysis) or sensors (Ludi *et al.* 1990) are common.

FIA has since become almost universally accepted as an important analytical tool. Samples can be introduced either continuously or as a well-defined plug in a flowing stream for transport to a detection device where the concentration of the analyte is measured. The sample can be processed in a number of ways in the flow system (e.g., dilution, chemical reaction, dialysis etc.) prior to detection. The operation can be performed automatically and reproducibly as the sample is transported

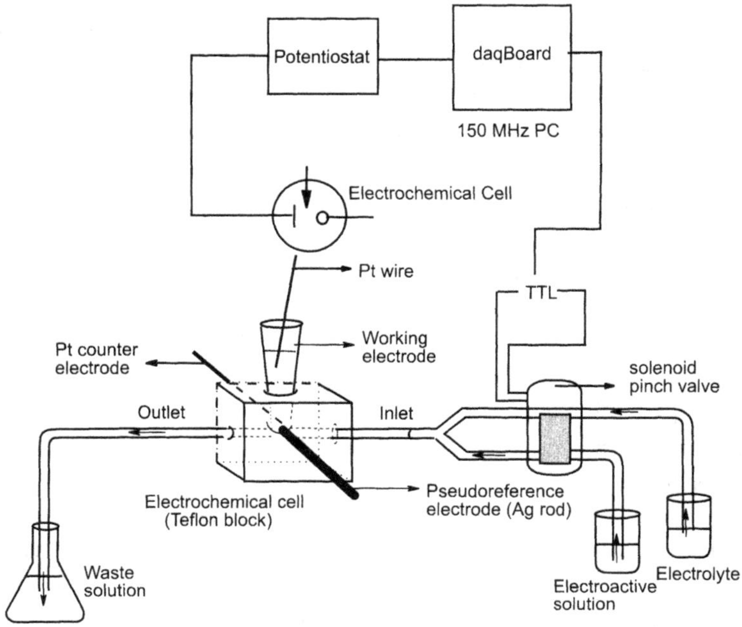

Figure 4. Schematic of a homemade flow-injection assembly designed in our laboratory. For instrumentation details see the text.

from the place of introduction to the place of detection. The carrier may be a homogeneous liquid, or it may be segmented by gas bubbles. The introduction of gas bubbles decreases the dispersion of the sample drastically and consequently decreases the time of analysis. The technique is widely exploited in many commercial auto-analyzer systems. Stewart *et al.* (1980) developed a microprocessor-controlled system for automated flow injection analysis. This control system is relatively inexpensive and is suitable for use by inexperienced personnel.

We have designed a simple set-up for carrying out FIA studies in our laboratory. It exhibits high sensitivity and ease of operation (Varma and Mitra 2002).

A small teflon block ($25 \times 25 \times 25$ mm^3) was used as an electrochemical cell. Holes were drilled at the centre of five faces that meet at the centre of the block. This arrangement provides a sufficiently small working volume. A schematic representation of our homemade flow-injection set-up is shown in Figure 4. The working electrode was made from a micropipette tip. It was packed with modified enzyme paste and was fitted without leak on the top face. The reference electrode (a silver rod) and the counter electrode (a platinum wire) were fixed with adhesive on the two opposite faces of the cube. The electrodes were connected to the potentiostat (homemade) in the standard way.

Tygon tubes were also fixed with adhesives on the other two opposite faces of the cube. They served as inlet and outlet. The inlet is bifurcated into two; through one of the inlet tubes, the electrolyte solution was allowed to flow through the chamber by gravity only (~2 ml/min). Into the other inlet tube, electroactive solution (glucose or H_2O_2 depending on the enzyme electrode) was allowed to flow under the control of a 12 V two-way solenoid pinch valve (Biochem valve) into the electrochemical cell, the outlet, and finally into the waste solution flask. The solenoid valve was interfaced with a transistor driver so that the opening and closing of the valve can be controlled by logic pulses (digital output of the daqBoard) from the computer. The current values were automatically collected from the I/E output of the potentiostat and stored in a file. All the necessary software for this purpose was written by us in Borland C++.

1.6.3. Multiple Sensors

One sample can be used for analysis or detection of several components (DNA arrays). In contrast to the above two configurations, in which only one sample is analyzed at a time, a multiple sensor set-up can be used in which a number of samples are analyzed simultaneously. DNA arrays are a common example (Millan and Mikkelsen 1993). These DNA microarrays typically consist of up to 20 000 individual spots, each of which contain millions of copies of a specific cDNA probe immobilized on a specially prepared support. Fluorescent-labeled targets, which contain a proportional representation of all of the transcripts present in a given tissue sample, are subsequently hybridized to the microarray. Generally the control and the experimental samples are labeled with two different fluorescent dyes, and then both targets are hybridized simultaneously to a single microarray. In this manner, genes whose transcript abundance changes in response to experimental treatment can be detected as spots, due to the variation of signals coming from the two dyes. The visualization is usually by optical method (fluorescence). It may not be in the very distant future when we shall have similar arrays using enzymes or antibodies.

1.7. Future Directions in Biosensors (Micro and Nano Technologies)

The expressions "nanosensors" and "microsensors" are gaining importance (Wiesendanger 1995). In a general way a nanosensor can be defined as a sensor fulfilling one of the three following requirements:

(i) The sensitivity of the sensor is in the nanoscale (e.g., sensors for displacements on the nanometer scale, force with the sensitivity of nanonewtons, sensors for the measurements of current of the order of nanoamperes, sensors for magnetic fields of the order of nanotesla and power sensors with the sensitivity of nanowatts).

(ii) The spatial confinement of the interaction of the sensor with the object is on nanoscale combined with the scanning technologies that directly leads to a spatially resolved sensor technology with a spatial resolution on nanometer length scale.

(iii) The size of the sensor is on nanometer scale.

The advancement in this field is mainly due to the progress in the field of silicon-based microfabrication and the influence of scan probe technology due to the invention of the scanning tunneling microscope (STM) by Binnig and Rohrer (1982, 1987).

In the first 10 years since its invention, STM and related scanning-probe-based techniques have mainly been used for imaging, measuring and modification tools (Guentherodt and Wiesendanger 1992). However, the impact of STM on sensor technology has become evident and an increasing number of nanosensors derived from scanning probe technology are being developed.

1.8. Immobilization and Types of Immobilization

Immobilization is the process in which a soluble fraction (usually a biomolecule) is rendered insoluble usually by attachment to a solid phase. The attachment can be as simple as adsorption or may be quite complex as in covalent coupling. Cross-linking is a special kind of immobilization in which the soluble molecules are made insoluble using some bifunctional chemical agent leading to polymerization. Immobilized enzyme preparations may be more effective as catalysts since they are recoverable and possibly more stable than free enzymes (Zaborsky 1973). Drawbacks of this technique include the heterogeneous nature of reaction and the ability of the catalyst to catalyze only a single reaction. A large number of bioreactors using immobilized enzymes are being routinely used commercially in biotechnology industry.

A number of methods of immobilization of biological molecules exist, and no single method is perfect for all molecules or purposes. Figure 5 schematically shows the various immobilization techniques. When attaching a biologically active molecule to a support, it is important to avoid a mode of attachment that reacts with or disturbs the active site(s) of the molecule, as loss of activity may result otherwise. It is

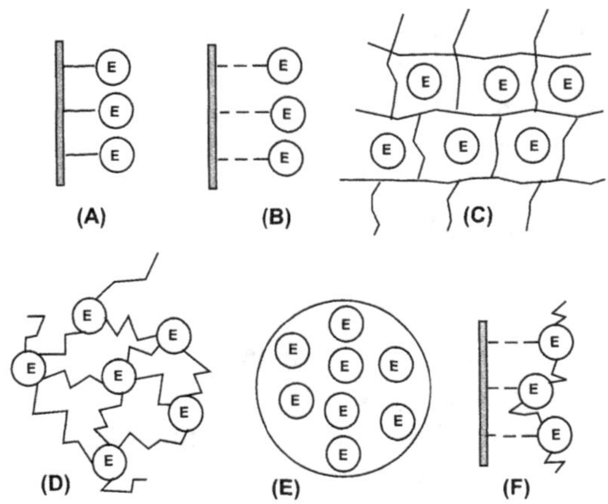

Figure 5. Schematic representation of various immobilization techniques: (A) covalent binding; (B) adsorption; (C) gel entrapment; (D) cross-linking; (E) microencapsulation; (F) adsorption-cross-linking.

also important to avoid overloading the matrix with the enzyme, since overloading leads to overcrowding and hence reduced activity. Particular attention to the way in which the macromolecule can be attached to the insoluble matrix and the choice of matrix is also a matter of importance (Kennedy and Cabral 1983a, Bowers 1986). Methods for the immobilization of enzymes, cells, antigens, antibodies, nucleic acids, antibiotics, etc. have already been reviewed by other workers (Kennedy and Cabral 1983a). Immobilization can be done on a support, or in a matrix (which may be a porous gel) or on a semipermeable membrane.

The *adsorption* of an enzyme onto an insoluble support is the simplest method of immobilization. The procedure consists of mixing enzyme and support matrix (without activation) under appropriate conditions and, following a period of incubation, separating the insoluble material from the soluble material by centrifugation or filtration. The major disadvantage of this method is that the enzyme is not firmly bound to the support, and can be easily leached out. Changes in experimental conditions such as pH, ionic strength, temperature and type of solvent can cause desorption of the enzyme from the support. Besides salt linkages, other weak binding forces (e.g., hydrogen bonds, van der Waals forces) are also involved in the adsorption of an enzyme to the support material.

Microencapsulation focuses upon maintaining the solution environment around the enzyme rather than upon maintaining the physical or chemical forces necessary for immobilization. The original solution containing the enzyme is wholly immobilized rather than selectively immobilizing the particular enzyme molecule. Microencapsulation creates artificial cells, which have a membrane similar to natural cells to control the size of molecules, which can move in or out of the cell. Large molecules such as enzymes or proteins can be retained within the encapsulated sphere, while small substrate and product molecules can freely diffuse across the synthetic membrane. One of the advantages of microencapsulation over regular entrapment of enzymes is the high surface area possible per unit of enzyme immobilized, allowing high effectiveness factors and high concentration of enzymes in the original solution. A great deal of current research is directed towards controlled drug release from microcapsules. (Chang 1977, Patwardhan and Das 1983). The diameters of the microspheres can range from a few to several thousand microns, and the membrane thickness can go from hundreds of Ångstroms to several microns (Madan 1979). It is possible to microencapsulate combinations of enzymes, cofactors and their enzymes for regeneration, proteins, whole cells, ion-exchange resins, activated carbon particles and magnetic particles. Even small microcapsules can be trapped between large microcapsules. However, microencapsulation does not find wide application in biosensors.

Covalent coupling for the immobilization of enzymes is based upon the formation of covalent bonds between the enzyme molecules and the support material. It is important that the amino acids essential to the catalytic activity of the enzyme are not involved in the covalent linkage to the support. This may sometimes be difficult to achieve, and enzymes immobilized in this fashion often lose activity upon immobilization. This may be prevented if the enzyme is immobilized in the presence of a suitable competitive inhibitor – a step that would tend to have a protective effect on the catalytic site of the enzyme during immobilization. The covalent linkage of an enzyme to a support material often requires that the support material be 'activated'. Activation of the support matrix is nothing more than introducing suitable chemical functional groups that can take part in subsequent steps in the immobilization process.

Our aim is to design a simple disposable low-cost biosensor with long shelf life using covalently coupled enzymes. Using third-generation techniques, we can design a biosensor that is not dependent on the oxygen present in solution nor does it use organic dyes or mediators. The formation of a covalent bond between the biocatalyst and the electrode creates a stable conjugate that is unlikely to dissociate during normal operation. In a covalently coupled system, the problem of enzyme leaching out often encountered in other immobilization methods (like adsorption) can be avoided. Covalent immobilization is promising because the closer proximity of the conjugate with the transducer improves the electron transfer rates. Immobilization on a conducting matrix (e.g., a metal electrode) is generally difficult but is desirable in biosensor application. Several enzyme biosensors have been designed and constructed using various kinds of immobilization techniques. Recent works dealing with chemically modified electrodes have made possible the permanent chemical modification of various electrode materials (Watkins *et al.* 1975). For example, gold electrodes have been covalently modified using thiol compounds (Katz *et al.* 1997). Glassy carbon is not an attractive material for enzyme immobilization as its chemical inertness towards coupling reagents is very high. Enzyme adsorptions on graphite or activated carbon followed by glutaraldehyde (Liu *et al.* 1975) or soluble carbodiimide (Cho and Bailey 1979) cross-linking have been reported. Recently covalent immobilization of the glucose oxidase enzyme onto graphite powder by carbodiimide treatment was reported (Gorton 1995). The direct covalent binding of the enzyme by carbodiimide coupling onto glassy carbon after initial activation by electrochemical oxidation was reported by Bourdillion (Bourdillion *et al.* 1980).

1.9. Designing a Simple Biosensor

We have covalently immobilized several enzymes onto the surface of glassy carbon-supporting matrix. The glassy carbon matrix is in the form of a fine powder (1 μm particle size), procured from Hochtemperatur Werke (Germany).

Several forms of sensors were studied; for instance potentiometric, amperometric enzyme electrodes were made as electrochemical cells. We have also devised electrodes as disposable strips with modified enzyme paste screen-printed onto the plastic surface. Recent developments in sensor technology have resulted in a variety of microsensors, manufactured using the sophisticated and economical techniques of microfabrication. Microfabricated sensors have electrical outputs that can be interfaced directly with computers and provide information for process development and characterization.

In order to make a third-generation biosensor we have separated the apoenzyme (i.e., the enzyme without the prosthetic group) from the glucose oxidase and reconstituted the native enzyme with the matrix on which FAD has been covalently coupled with a sufficiently long spacer arm. We find the reconstituted enzymes show more activity when immobilized than their native counterparts. Similar experiments have been performed with horseradish peroxidase, and a better electron transfer was observed. The electrode at this stage can be stored in the refrigerator. We have also tried other immobilization schemes using divinylsulfones to get higher enzyme loading and to reduce the number of steps using harsh treatments.

2. METHODOLOGY

2.1. Covalent Immobilization Protocol

The covalent immobilization of the protein (enzyme) onto the glassy carbon surface includes three major steps:
(1) Activation.
(2) Addition of spacer arm.
(3) Enzyme coupling.

2.1.1. Activation

2.1.1.1. Heat Treatment

The glassy carbon matrix was heated in a hot air oven at 110–120°C for 2–3 hours. This helps in the removal of volatiles and other adsorbents from the glassy carbon particles, thereby increasing the surface activity. Washing the matrix with organic solvents like CCl_4 also helps in removing the tar and other non-volatile substances, which may be present on the carbon surface from the manufacturing process.

2.1.1.2. Activation

We have followed the procedure described below to introduce functional groups onto the glassy carbon surface.

Hydroxylation: To 400 mg of the preheated matrix a mixture (10 ml) of concentrated H_2SO_4 and H_2O_2 (3:1) was added and left for 24 hours at room temperature. This treatment makes the matrix hydrophilic by introducing –OH groups on the surface glassy carbon powder. The mixture was heated in hot water bath for 6 hours to complete the reaction and to decompose any residual hydrogen peroxide. The excess acid was neutralized with a frozen solution of 1N NaOH and the matrix washed with 0.1% NaCl by centrifuging at 8000 rpm for 15 minutes at least thrice. The supernatant was checked for neutrality after every wash. The carbon matrix sample was collected at the end. A part of this was used as a control. All the matrices were first washed twice with 10–15 ml of doubled distilled water, then washed twice with 0.1% NaCl followed by a final wash with a mixture of methanol and DDW (1:1). All washings were done by spinning the matrix at 8000 g for 10–15 minutes. The final product was dried in a vacuum desiccator. The supernatant was discarded and the pellet processed for the next step.

2.1.2. Addition of Spacer Arm

To the activated matrix a 16 atom spacer arm was attached in three steps as follows.

2.1.2.1. Carboxymethylation

Carboxymethylation of the matrix was carried out by stirring 350 mg of the matrix with a mixture of 50 ml of DMF (dimethylformamide) and 100 mg of K_2CO_3 for 6 hours under anhydrous condition. About 100 mg of chloroacetic acid is added to the above mixture and is stirred for 4 days at room temperature on a magnetic stirrer. This is then centrifuged at 8000 g for 15 min. The residue is then washed with distilled water to remove excess alkali, suspended in methanol to remove H_2O, and dried in a desiccator. This treatment introduces $-CH_2COOH$ functional groups onto the matrix surface. Chloroacetic acid in the presence of alkali and water forms glycolic acid. This reaction predominates more in the aqueous solution. In order to reduce the production of OH^- ions, an organic solvent (DMF) and a weak base like potassium carbonate (K_2CO_3) were used.

2.1.2.2. Hexamethylenediamine Coupling

In this step a support matrix bearing primary amino group was prepared by the reaction of the carboxymethylated matrix with hexamethylene-diamine in the presence of EDC [1-ethyl-3-(3-dimethylaminopropyl)-carbodiimide hydrochloride] as the condensing agent. This is done by incubating the matrix at 4°C for 24 hours with a mixture of 45 mg of carbodiimide and 20 mg of hexamethylenediamine dissolved in 5 ml of 0.03M H_3PO_4 (pH 4). The matrix is next washed thrice with 0.1% NaCl. The carboxymethyl groups (with the help of carbodiimide ($R_1-NH=C=NH-R_2$)) couple to one end of hexamethylenediamine ($CH_3-(CH_2)_6-CH_3$) (the other end containing an $-NH_2$ (amino) group will remain free). Hexamethylenediamine is employed in this step in order to promote a one-ended reaction; a single amide link is formed leaving a free amino group at the other end. At pH 4, the coupling of the proximal amino group to the matrix is considerably enhanced, with respect to the amino group remaining after attachment. Use of HMDA provides a longer spacer arm for the enzyme.

2.1.2.3. Glutaraldehyde Coupling

15 ml of 2.5% glutaraldehyde in 0.05M Na_2HPO_4 buffer (pH 7.0), is added to 300 mg of amino matrix, which is washed thrice with water after incubation for 1 hour at room temperature. In this step, the free amino end of hexamethylenediamine (i.e., the amino matrix) is coupled to one end of glutaraldehyde and the other end of the molecule with a –CHO group remains free. We thus obtain a glassy carbon surface with a 16-atom linker containing a free aldehyde group at the other end to which the enzyme (protein) can be coupled. The length of the spacer arm plays a key role in enzyme activity. If the spacer arm is too long, the rate of electron transfer to the matrix is expected to be reduced. On the other hand, an enzyme very close to the matrix has too little conformational freedom and poor access to the substrate and

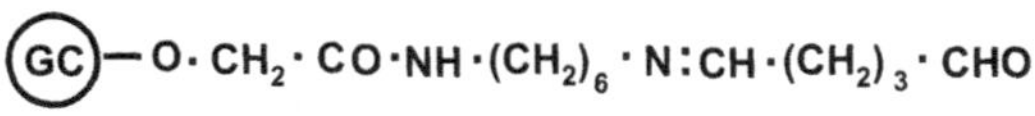

Figure 6. 16-atom spacer arm attached to glassy carbon particle.

its (active site's) environment may get dehydrated, leading to reduced activity. Therefore, a careful and judicious choice has to be made while deciding the length of the spacer arm.

2.1.3. Enzyme Coupling

To about 300 mg of the aldehyde matrix (Figure 6), 1 mg of protein in 5 ml of 0.05 M phosphate buffer was added and the coupling was allowed to take place for 3–4 hours. As before, this mixture was first washed twice by addition of 10–15 ml of doubled distilled water, then washed twice with 0.1% NaCl followed by a final wash with an 1:1 (v/v) mixture of methanol and DDW, at 8000 g for 15 min. The supernatant was discarded and the pellet dried in a desiccator.

2.1.4. Estimation of Protein Coupled to the Matrix

In order to determine the amount of protein coupled to the matrix we carried out standard methods of protein estimation and activity assay (depending upon the enzyme coupled). In the next step alkaline hydrolysis was carried out to separate the protein from the matrix, and the supernatant containing the protein was estimated.

2.1.5. Electrochemical Set-up

We have used several types of set-up.

2.1.5.1. Electrochemical Cell

We have used 10 ml working volume electrochemical cell with standard three electrodes. The working electrode was a glass tube with Pt wires inserted in one of the ends on the surface of which the modified paste was packed. The reference electrode was Ag/AgCl and the counter electrode was coiled Pt wire. A typical electrochemical cell can be

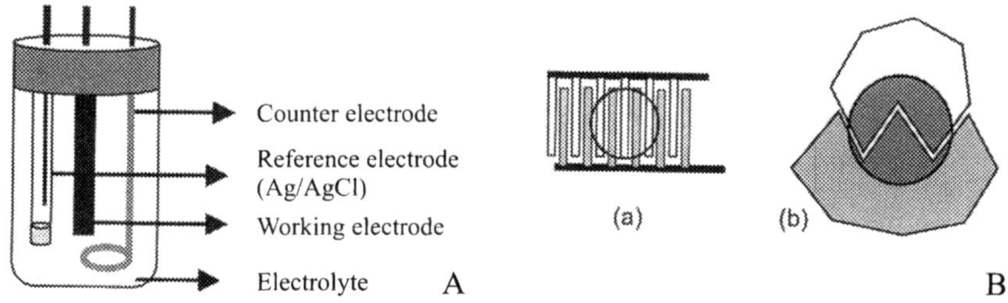

Figure 7. (A) Typical electrochemical cell, with standard three-electrode set-up comprising of working, counter and reference electrodes. (B) Simple patterns commonly employed for designing disposable strip electrode. The two electrodes are seen with different shadings and are placed within 1 mm of each other. One of the electrodes is covered with the enzyme ink. The shaded circle is a mask (the solution contacts the electrodes in this region only). We have used pattern (a) to make disposable strips in our work.

seen in Figure 7A. We have used this configuration in order to prepare microelectrodes for amperometric studies.

2.1.5.2. Disposable Strips

We have also made disposable strips by printing the enzyme modified paste onto a polystyrene (transparency sheet). This was done in two steps:

(i) *Ink preparation*: In order to print the electrodes we have prepared ink, which is a combination of an enzyme-modified matrix, binder, detergent and a solvent. Different detergents, sorbitan monostearate, sorbitan monolaurate, Triton X, Tween 40 and Tween 60 have been tried out and finally Tween 40 was selected as it was found to disperse the glassy carbon particles uniformly. The binder was prepared by mixing different proportions of PVA and PVP. A final composition of 4:1 PVP:PVA in water was standardized. We used water as a solvent for making the ink and GOD was the model enzyme. The final composition of the ink is given in Table 3.

(ii) *Printing the electrodes*: Polyester film (overhead transparency) was used as the plastic substrate to print the conducting base and the counter electrode. A 7475A Hewlett Packard plotter and Staedler 1.0 mm pens were used to print the electrodes and conducting base. At a time, about 70 strips can be made with this device.

The disposable strip electrode is prepared in three steps:

Table 3. Components of ink

Component	Amount
Glassy carbon enzyme matrix	125 mg
Water	150 μL
Tween 40	20 μL
Binder (PVP+PVA)	90 μL

(1) The design is drawn with silver paint using a plotter. The silver paint serves as a reference electrode.
(2) Modified carbon ink (*vide supra*) is plotted over it, using the same device. This serves as the working electrode. A thin silver connectivity was used in order to have an electrical contact.
(3) The active region is protected by dipping in 2% cellulose acetate solution.

The silver contacts are kept exposed for electrical connectivity. The adherence of the paste to the polystyrene support is good and showed no signs of peeling. An essential feature of the biosensor is the interface between the biochemically sensitive coating and the transducer. An intimate contact is achieved by immobilizing the biocatalyst at the surface of the transducer. The substrate to be determined diffuses through the membrane into the enzyme layer where the enzymatic reaction occurs.

When a molecular wire is used for connecting the electrode (glassy carbon particle) and the enzyme (preferably to the active site) the electrical connection between the glassy carbon particle and the base substrate (a silver electrode drawn with a silver paint) is most important. In order to apply the sample and connect the strip to the electrochemical analyzer in a convenient way different designs have been tried out to fabricate the strips. Some of them are given in Figure 7B.

3. RESULTS AND DISCUSSION

Although there has been considerable interest in the electrochemical behaviour of GOD (glucose oxidase), only few reports are available in the literature directly related to the interaction between GOD and electrodes. Aided with sensitive techniques such as differential pulse voltammetry, it has been shown that GOD undergoes direct electron exchange on mercury

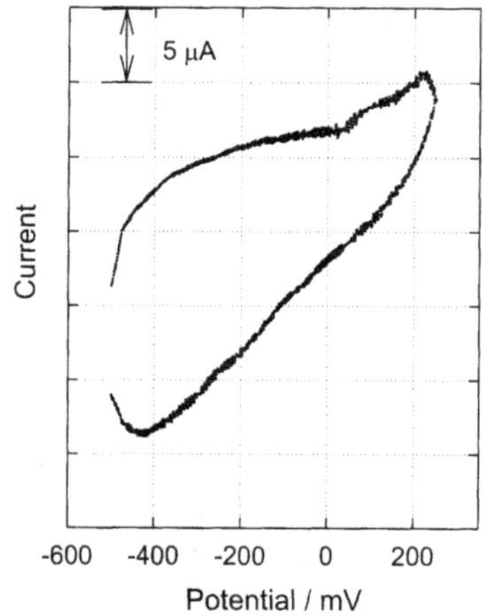

Figure 8. CV of enzyme electrode with no adsorbed mediator species. The enzyme was immobilised using a long spacer arm to the glassy carbon powder. The electrode was used in a paste form. The potential scan was taken in presence of glucose (1 mM). Two small irreversible peaks can be observed that correspond to the reduction of the enzyme-bound FAD cofactor. Scan rate: 0.5 mV/s.

or graphite (Narasimhan and Wingard 1986) or platinum (Scheller *et al.* 1979) electrodes. We have studied the mechanism of electron transfer in paste electrode systems. In our work the role of mediator was found to be central for an efficient electrochemical contact with electrode surface and a mechanism has been proposed. We have also observed that the FAD cofactor bound to glucose oxidase enzyme aids in electron transfer to the electrode surface that was found to be irreversible as observed by cyclic voltammetry experiments. Further the electrode surface reactions in presence of oxygen (dissolved) were found to be electrochemically different (hydrogen peroxide gives a reduction current) from the reactions in the absence of oxygen (oxidation current).

Mediated electron transfer from enzyme active site to the electrode: The cyclic voltammogram of the enzyme electrode (without mediators) in the presence of the substrate (glucose) shows a single clear irreversible peak as seen in Figure 8 at a potential ~100 mV (vs. SCE) although a second small peak can be seen around ~225 mV. These peaks are due to the buried FAD, which undergoes 2e $2H^+$ redox reactions. The peak potentials are considerably shifted from native FAD immobilised (Deverakonda and Mitra 1998), due to binding with the enzyme. The electron exchange also becomes irreversible in the holoenzyme as seen in Figure 8.

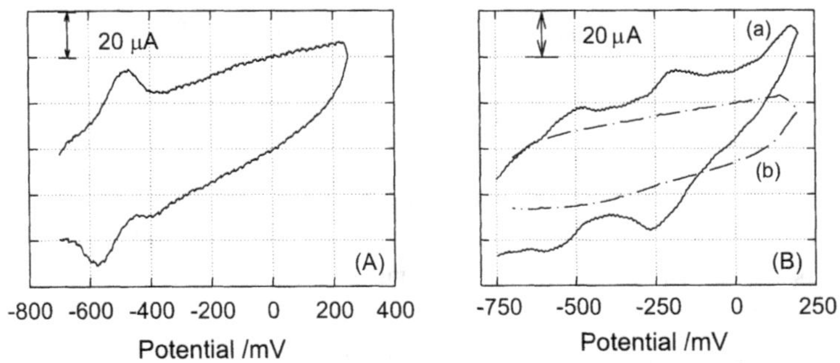

Figure 9. (A) CV of FAD in solution in the presence of a blank electrode surface. A well-defined peak centered at −590 mV was observed upon reduction of FAD. In the return sweep, a peak was seen at −465 mV, corresponding to its oxidation. (B) Cyclic voltammogram of FAD covalently attached to glassy carbon matrix by a 13-carbon long spacer arm. Reduction peaks are observed at −260 mV and −601 mV. The return sweep shows one peak at −475 mV and another sharp peak at −197 mV. It seems that the peaks represent two redox couples. Scan rate: 10 mV/s; supporting electrolyte 100 mM KCl in 10 mM phosphate buffer.

The CV of the enzyme electrode with adsorbed quinone in presence of glucose shows a clear peak due to quinone at a potential of about −100 mV (vs. SCE). This clearly shows that in the presence of the second mediator (riboflavin), there is a shift in the potential of quinone (~70 mV), suggesting a strong interaction between these two mediators. The peak due to the bound FAD (enzyme) also appears considerably sharper due to rapid electron exchange in presence of mediators. We also observe that electrochemical detection is optimum in the potential window of −300 to −100 mV, close to the quinone peak.

The cyclic voltammetric behavior of FAD in solution in presence of blank paste electrode is shown in Figure 9A. A well-defined peak centered at −590 mV was observed upon reduction of FAD. In the return sweep, a peak was seen at −465 mV, corresponding to its oxidation. Earlier reports (Scheller *et al.* 1979) have established that electrochemical reduction of FAD takes place by two indistinguishable one-electron processes. Adsorption of FAD on the electrode surface leads to its reduction at more cathodic potential (11). Figure 9B shows a cyclic voltammogram of FAD covalently attached to glassy carbon matrix by a 13-carbon-long spacer arm. Reduction peaks are observed at −260 mV and −601 mV. The return sweep shows one peak at −475 mV and another sharp peak at −197 mV. It seems that the peaks represent two redox couples. The matrix with the linker alone does not show any of these characteristics, suggesting that the redox center lies on the FAD molecule. Reduction peaks at −350 mV and −680 mV have been reported (Wingard and Gurecka 1980) for the covalent attachment of FAD onto a solid glassy carbon electrode. In our case we observed that reduction peaks are well separated from the oxidation peaks, and the immobilized FAD shows relatively more reversible behavior compared to the free FAD in solution.

FAD is known to undergo redox reaction where two protons and two electrons are released or taken up. In glucose oxidase, FAD is deeply seated in a cavity and therefore is not easily accessible for conduction of electrons to the electrode surface. Immobilization of FAD onto the matrix by a long spacer arm leads to the conduction of electrons from the FAD molecule to the electrode surface. A long spacer arm is expected to facilitate binding of the cofactor (FAD) to the apoenzyme. When GOD enzyme is reconstituted by adding the apoenzyme of GOD to the FAD immobilized matrix, it facilitates easy electron transfer from the enzyme redox center to the electrode surface and the response is sharper.

When the electrode is prepared with a paste that contained riboflavin or flavin mononucleotide (mixed with the paste during preparation), a small (50–100%) enhancement of the signal was actually observed. In this case, it is probably acting as a mediator (that can exchange electrons between the active site and the electrode surface). However, the active site in GOD is deeply seated and interaction with the mediator is not efficient enough. When the molecule (FAD) is covalently attached, there may be a far more efficient route for the electron transfer between the electrode and the active site. A rate enhancement of ~1000 times was observed in our experiments. It is perhaps possible to get still higher efficiencies if the properties of the molecular wire can be suitably optimized (e.g., by introducing alternate double-bonded chains in the spacer arm). This remains a possibility that needs to be further explored.

There has been a rapid development in biosensor design and associated instrumentation in recent times. Recent advances in silicon (semiconductor) technology and polymer chemistry have revealed new materials and methods available for exploitation as sensors. Of all the signal transduction methods, electrochemical transduction has been the most widely used and favored method in both research and industry. Even in electrochemical methods, amperometry is the best configuration owing to the simple set-up and interface. The advent of microelectrodes has led to the concept of internal signal processing that originates from the development of chemically sensitive field effect transistors.

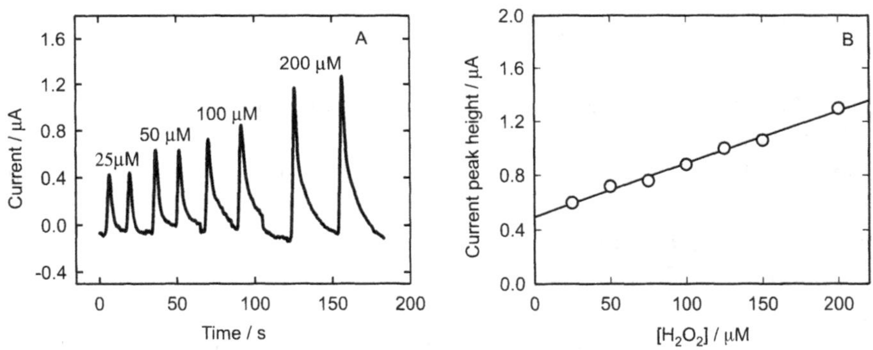

Figure 10. (A) FIA response of HRP-modified paste electrode to H_2O_2. The base of the peaks shows broadening with the increase in the concentration of the substrate, indicating a sluggish electron transfer. (B) Amperometric response of the HRP electrode, showing a clear linear increase in the current values with the concentration of the substrate.

A rapid test for clinically important enzymes like amylase, creatine kinase, amino acid transferase, etc. would be invaluable in emergencies while the biochemical measurement of substances such as lactate, glucose, urea, creatinine, etc. makes continuous monitoring at the bedside feasible. Biosensors specific for these analytes would serve the purpose. Thermistors can be used to measure very small temperature changes that take place as reactions proceed. Linking such transducers to appropriate enzymes allows biosensors to be constructed that are responsive to a variety of biologically important molecules. However, a majority of researchers in both academic and industry has concentrated their developmental efforts on one goal – commercialization of a portable glucose biosensor for use by diabetics.

Similar studies on other enzymes, e.g. horseradish peroxidase (HRP), have been carried out. Figure 10 shows the FIA response of native HRP paste electrode. We have used the indigenously made set-up shown in Figure 4 (vide supra). An optimum bias of +50 mV was used. The peaks are very sharp and clear, each peak corresponding to a single injection of ~300 μL of H_2O_2. The base of the peak increases with the concentration. A tailing effect is observed as the concentration of the substrate increases, suggesting that the electrode processes are slow. If the redox center of the enzyme (hemin) is bound directly to the electrode with a molecular wire a better electron transfer kinetics would be observed as seen in the FIA (Varma 2002) and electrochemical impedance spectroscopy (EIS) (Varma and Mitra 2002) techniques.

4. CONCLUSION

The constant search for simple and reagentless methods for specific determination of various metabolites is still an important goal for numerous laboratories working in the attractive field of biosensors. The use of mediated electron transfer for shuttling electrons between electrode and enzyme is becoming popular in glucose sensors. Approximately 1 billion Exatech strips were sold last year. Even if the design of associated electronics and software for signal processing appear less critical, immobilization techniques, stability of bioreactive components and improvement of signal transduction at the enzyme–transducer interface still appear as the major bottlenecks for such systems. 'Protein engineering' with site-directed mutagenesis that allows amino acid substitution will lead to a better understanding of protein stability and thus to the production of more stable biocatalysts.

People are increasingly concerned about the contents of the food they eat. There is a growing need to be informed about the levels of additives, growth hormones and cholesterol in food. Contamination in the food chain, due to pesticides or for instance the presence of *Salmonella* in the poultry and egg industry, is increasing the need for faster and more sensitive testing instrumentation. In addition, the consumer demand for fresh food and the industry's desire to reduce spoilage makes food testing one of the more interesting biosensor application fields.

Other issues under investigation include devices for monitoring lactate, SO_2 and aldehyde, the alternative for monitoring rancidity of oils and fats. Cranfield Institute of Technology has patented the concept of the organic phase enzyme electrode, which can be used to detect peroxides in organic solvents and may be useful in detecting rancidity in oils and fats.

If we compare the present biosensors with the natural ones (for example, the nose or the eye), they are very crude and simplistic. The recognition molecules in the 'natural sensors' are not necessarily highly specific but the signal transduction via the biomolecules is sophisticated. The specificity often comes from processing of the data collected and recognizing the pattern via a continuous learning process. This mode of operation using the data collected from multiple biosensors is expected to be exploited in the future because the ever-increasing capability of microprocessors will provide fast computation.

Third-generation biosensors have built-in signal processing circuitry. When such sensors are combined with the microscale valves and actuators currently under development (utilizing micromachining technology), a whole analytical instrument can be built on a silicon wafer. Such an instrument can be mass-produced and used in a variety of applications including homes, hospitals, automobiles, toxic-dump sites, etc.

The effort to continuously increase the density of electronic components to obtain smaller 'packages' will be limited eventually, not by the microlithographic technique employed but by the minimum size allowable for a transistor. Many biological molecules are able to synthesize complex self-organizing molecules with apparently just the required electronic properties. This suggests that the solution to this problem may be found in replacing silicon with biomolecular components. This idea has led to the proposition of many *molecular-electronic* systems. In the past, materials and processing methods developed for microelectronic applications have been exploited in sensor developments. Therefore, any future developments in molecular electronics are expected to be important for biosensor technology.

There are several potential applications of biosensors in industrial fermentation processes for monitoring substrate levels, product formation and microbial biomass. The development of biosensors for fermentation is however hampered by the need of many food and pharmaceutical formulations for steam sterilization to prevent microbial contamination.

There is an increasing demand for inexpensive and reliable sensors to allow not only routine monitoring in the central or satellite laboratory, but also analysis with greater patient contact, such as in the hospital ward, emergency rooms, and operating rooms. Ultimately, patients themselves should be able to use biosensors in the monitoring and control of some treatable condition, such as diabetes. It is probably true to say that the major biosensor market may be found where an immediate assay is required. If the cost of laboratory maintenance is counted with the direct analytical costs, then low-cost biosensor devices can be desirable in the whole spectrum of analytical applications from hospital to home.

At the present time, it must be noted that there is a shortage of efficient sensors at moderate prices that could promote rapid innovations of product and processes with the aid of microtechniques and microelectronics. The availability of low-cost sensors plays an important role in ensuring that the implementation of these systems will be marketable, especially when large numbers of units are involved. This goal is usually achieved only via "batch capable" microtechniques. Some of these technologies for miniaturization are already available. At this point, the great potential of sensors now on the market must also be addressed, because here also the evolution of sensor techniques is going in the direction of miniaturization, multi-functionality, integration and intelligence.

ACKNOWLEDGEMENT

Work reported in this chapter has been made possible by grants from the Board of Studies in Nuclear Sciences (DAE), Department of Science and Technology (DST) and the Council of Scientific and Industrial Research (CSIR).

REFERENCES

Bartlett, P.N., R.G. Whitaker, M.J. Green and J. Frews (1987). Covalent binding of electron relays to glucose oxidase. *J. Chem. Soc. Chem. Commun.*, **103**, 1603–1604.

Binnig, G., and H. Rohrer (1982). Scanning tunneling microscopy. *Helv. Phys. Acta*, **55**, 726–735.

Binnig, G., and H. Rohrer (1987). Birth and infancy of scanning electron microscopy. *Phys. Bl.*, **43**, 282–290.

Bourdillion, C., J.P. Bourgeois and D. Thomas (1980). Covalent linkage of glucose oxidase on modified glassy carbon electrodes – kinetic phenomena. *J. Am. Chem. Soc.*, **102**, 4231–4235.

Bowers, L.D. (1986). Applications of immobilized biocatalysts in chemical analysis. *Anal. Chem.*, **58**, 513–530.

Broyles, C.A., and G.A. Rechnitz (1986). Drug antibody measurement by homogenous enzyme immunoassay with amperometeric detection. *Anal. Chem.*, **58**, 1241–1245.

Carr, P.W., and L.D. Browers (1980). *Immobilized Enzymes in Analytical and Clinical Chemistry*. Wiley, New York, p. 197.

Cass, A.E.G., G. Davis, G.D. Francis, H.A.O. Hill, W.J. Aston, I.J. Higgins, E.V. Plotkin, L.D.L. Scott and A.P.F. Turner (1984). *Anal. Chem.*, **56**, 667–671.

Chang, T.M.S. (1977). In *Biomedical Applications of Immobilized Enzymes and Proteins*. T. Chang (Ed.), Plenum Press, New York, p. 69.

Cho, Y.K., and J.E. Bailey (1979). Immobilization of enzymes on activated carbon: selection and preparation of carbon support. *Biotechnol. Bioeng.*, **21**, 461–476.

Clark, L., and C. Lyons (1962). Electrode systems for continuous monitoring in cardiovascular surgery. *Ann. N.Y. Acad. Sci.*, **102**, 29–45.

Claycomb, R.W., and M.J. Delwiche (1998). Biosensors for on-line measurement of bovine progesterone during milking. *Biosensors Bioelectron.*, **13**, 1173–1180.

Corcoran, C.A., and G.A. Rechnitz (1985). Cell based biosensors. *Trends Biotechnol.*, **3**, 92–96.

Deverakonda, S., and C.K. Mitra (1998). Electrochemistry of reconstituted glucose oxidase on carbon paste electrodes. *Bioelectrochem. Bioenerg.*, **47**, 67–73.

Frew, J.E., and H.A.O. Hill (1987). Electrochemical biosensors. *Anal. Chem.*, **59**, 933–944.

Gorton, L. (1995). Carbon paste electrodes modified with enzymes, tissues and cells. *Electroanalysis*, **7**, 23–45.

Gough, D.A., J.Y. Lucisano and P.H.S. Tse (1985). Two-dimensional enzyme electrode sensors for glucose. *Anal. Chem.*, **57**, 2351–2357.

Guentherodt, H.-J., and R. Wiesendanger (Eds.) (1992). *Scanning Tunneling Microscopy I*. Springer, Berlin.

Guilbault, G. (1984). *Handbook of Immobilized Enzymes*. Marcel Dekker, New York.

Guilbault, G.G., and G. Lubrano (1973). Enzyme electrode for the amperometric determination of glucose. *Anal. Chim. Acta*, **64**, 439–455.

Guilbault, G.G., and J.G. Montalvo Jr (1969). Urea-specific enzyme electrode. *J. Am. Chem. Soc.*, **98**, 2164–2165.

Janchen, M., F. Scheller and D. Pfeiffer (1989). In *Biosensors: Applications in Medicine, Environmental Monitoring and Process Control*, GBF monograph. R.D. Schmidt and F. Scheller (Eds.), Verlag Chemie, Weinheim, pp. 17–25.

Katz, E., V. Heleg-Shabtai, B. Willner, I. Willner and A.F. Buckmann (1997). Electrical contact of redox enzymes with electrodes: novel approaches for amperometric biosensors; *Bioelectrochem. Bioenerg.*, **42**, 95–104.

Kennedy, J.F., and J.M.S. Cabral (1983a). In *Solid-Phase Biochemistry: Analytical and Synthetic Aspects*. W.H. Scouten (Ed.), Wiley, New York, p. 253.

Kennedy, J.F., and J.M.S. Cabral (1983b). In *Applied Biochemistry and Bioengineering*, Vol. 4. I. Chilbata and L.B. Wingard (Eds.), Academic Press, New York, p. 190.

Kobos, R.K. (1980). *Ion-selective electrodes in analytical and clinical chemistry*. Wiley, New York, p. 197.

Liu, C.C., E.J. Lahoda, R.T. Galasco and R.B. Wingard (1975). Immobilization of lactase on carbon. *Biotechnol. Bioeng.*, **17**, 1695–1696.

Ludi, H., M.B. Garn, P. Bataillard and H.M.J. Widemer (1990). *Biotechnology*, **14**, 71.

Madan, P. (1979). In *Microencapsulation – New Techniques and Applications*. T. Kondo (Ed.), Techno, Inc., Tokyo, p. 11.

Meixner, H. (1995). In *Sensors: A comprehensive survey*, Vol. 8. W. Gopel, J. Hesse and J.N. Zemel (Eds.), VCH Verlagsgesellschaft, Weinheim, Germany/VCH Publishers, New York, p. 1.

Millan, K.M., and S.R. Mikkelsen (1993). Sequence-selective biosensor for DNA based electroactive hybridisation indicators. *Anal. Chem.*, **65**, 2317–2323.

Narasimhan, K., and L.B. Wingard (1986). Enhanced direct electron transport with glucose oxidase immobilised on (aminophenyl)boronic acid modified glassy carbon electrode. *Anal. Chem.*, **58**, 2984–2987.

Patwardhan, S.A., and K. Das (1983). In *Controlled-Release Technology, Bioengineering Aspects*, K. Das (Ed.), Wiley, New York, p. 122.

Persaud, K., and G. Dodd (1982). Analysis of discrimination mechanisms in mammalian olfactory system using a model nose. *Nature*, **299**, 352–355.

Petersson, B.A., H.B. Anderson and E.H. Hansen (1987). Determination of urea in undiluted blood samples by flow injection analysis using optosensing. *Anal. Lett.*, **20**, 1977–1974.

Robinson, G.A., H.A.O. Hill, R.D. Philo, J.M. Gear, S.J. Rattle and G.C. Forrest (1985). Bioelectrochemical enzyme immunoassay of human chorionogonadotropin with magnetic electrodes. *Clin. Chem.*, **31**, 1449–1452.

Robinson, G.A., V.M. Cole, S.J. Rattle and G.C. Forrest (1986). Bioelectrochemical immunoassay for human chorionic gonadotrophin in serum using an electrode-immobilised capture antibody. *Biosensors*, **2**, 45–57.

Scheller, F., G. Strand, B. Neumann, M. Kuhn and W. Ostrowski (1979). Polarographic reduction of prosthetic group in flavoprotein. *Bioelectrochem. Bioenerg.*, **6**, 117–122.

Schubert, F., D. Pfeiffer, U. Wollenberger, S. Kuhnel, G. Hanke, B. Hauptmann and F. Scheller (1989). In *Biosensors: Applications in medicine, environmental monitoring and process control*, GBF monograph. R.D. Schmidt and F. Scheller (Eds.), Verlag Chemie, Weinheim, pp. 11–15.

Stewart, K.K., K.K. Brown and B.M. Golden (1980). A microprocessor control system for automated multiple flow injection analysis. *Anal. Chim. Acta.*, **114**, 119–127.

Thompson, M., and U.J. Krull (1991). Biosensors and transduction of molecular recognition. *Anal. Chem.*, **63**, 393–405.

Updike, S.J., and G.P. Hicks (1967). The enzyme electrode. *Nature*, **214**, 986–988.

Varma, S. (2002). Electrochemical studies on reconstituted horseradish peroxidase modified carbon paste electrodes. *Bioelectrochem.*, **56**, 107–111.

Varma, S., and C.K. Mitra (2002). Bioelectrochemical studies on catalase modified glassy carbon paste electrodes. *Electrochem. Commun.*, **4**, 151–157.

Varma, S., and C.K. Mitra (2003). Low frequency impedance studies on covalently modified glassy carbon paste. *Electroanalysis*, in press.

Watkins, B.F., J.R. Behling, E. Kariv and L.L. Miller (1975). A chiral electrode. *J. Am. Chem. Soc.*, **97**, 3549–3550.

White, J., S. Mall and J.S. Kauer (2002). Using biology to guide development of an artificial olfactory system, In *Neurotechnology for Biomimetic Robots*. J. Ayers, J. Davis and A. Rudolf (Eds.), MIT Press, Cambridge, MA, in press.

Wiesendanger, ?. (1995). In *Sensors: A comprehensive survey*, Vol. 8. W. Gopel, J. Hesse and J.N. Zemel (Eds.), VCH Verlagsgesellschaft, Weinheim, Germany/VCH Publishers, New York, pp. 338–363.

Wilson, K., and J. Walker (1996). *Principles and Techniques of Practical Biochemistry*, 4th edition, p. 571.

Wingard Jr, L.B., and J.L. Gurecka Jr (1980). Modification of riboflavin for coupling to glassy carbon. *J. Mol. Catal.*, **9**, 209–217.

Zaborsky, O.R. (1973). *Immobilized Enzymes*, CRC Press, Cleveland. OH, p. 177.

Chapter 2

OPTICAL DIAGNOSTICS FOR MEDICINE

N.K. Chaudhury

OUTLINE

Advances in Biosensors
Volume 5, pages 37–62.

1. INTRODUCTION

The development of optical diagnostics for medicine is an active area of research in biophotonics. This is because various bioanalytes of clinical importance have definite optical spectroscopic properties viz. absorption and fluorescence due to the presence of characteristic molecules. These analytes can be glucose, cholesterol, myoglobin, uric acid, lactate, therapeutic drugs, toxins, etc. The spectroscopic properties are due to the presence of characteristic chromophores in these biomolecules. Specific marker bands from these spectra can be identified and used for detection, characterisation and quantitation of biomolecules. These biomolecules can provide molecular level recognition for the analyte. The lowest concentration of a molecule that could be detected using absorption spectroscopy is about 10^{-6} M under optimum conditions while the limit for fluorescence detection can be as low as 10^{-15} M. Moreover, the fluorescence signal is relatively well defined, easier to detect and quantitate even in the presence of other molecules. For these reasons, fluorescence detection is a preferred method for quantitation. Although many biomolecules show characteristic fluorescence emission bands other bioanalytes are either poorly fluorescent or almost non-fluorescent in nature. A number of fluorescent dyes have been used to probe biomolecules ranging from nucleic acids, proteins, lipids, to drugs, etc. Fluorescent-labeled specific enzymes or antibodies can bind with the target molecule or analyte and the characteristic change in fluorescence intensity from the bound complex can be detected by using suitable opto-electronic devices such as a spectrofluorimeter. Accurate determination of the output signal (fluorescence intensity) is necessary in order to obtain a direct correlation with the amount of analyte present in the medium. The usual problem is the amount of amplification of the signal from the optical sensor required for acceptable electrical output. However, as a result of recent advances in technology associated with opto-electronic devices it is now possible to detect fluorescence from a single molecule with negligible interference (Johnson *et al.* 1989).

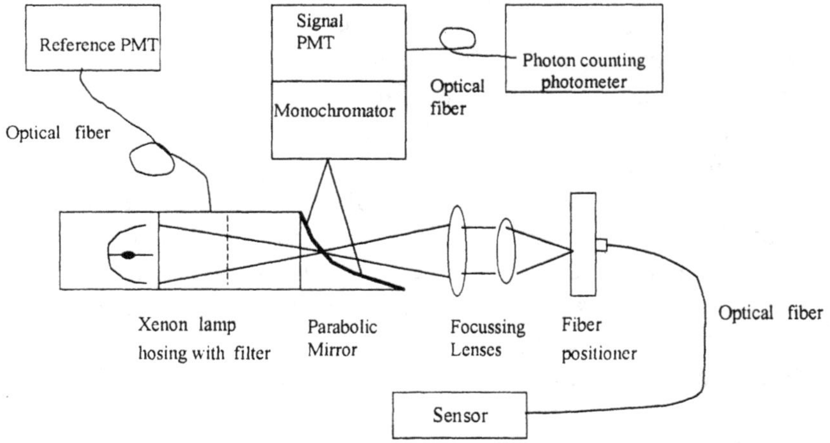

Figure 1. Typical set-up showing the major components of an optical biosensor.

Development of optical biosensors for clinically important bio-analytes requires that the sensor device (transducer), e.g., labeled enzyme/antibodies be interfaced with the surface of an optical fibre. Although techniques for immobilizing enzymes on solid surface have been known since 1970, the development of a sensor system for optical detection is still the major problem (Guilbault 1972). Fibre-optic biosensors can be classified as intrinsic or extrinsic depending upon the geometry of sensor. For example, an intrinsic sensor will require immobilization of analyte sensitive reagent, i.e., fluorescent-labeled antibodies, on the surface of the optical fibre and usually at the fibre tip. Extrinsic sensors consist of bare optical fibre dipped into the sample to which the fluorescent-labeled reagent has been added. In the latter case, the fibre-optic both carries the excitation light and collects the scattered signal. The instrumentation necessary for detection and quantitation is relatively easier to build as the basic methods are based either on spectrophotometric or fluorimetric techniques and all the necessary components are available as spares from multiple commercial sources. The basic components of an optical biosensor device would be a light source such as a low-power xenon or mercury lamp or low-power diode lasers along with a set of filters or monochromator for excitation of the fluorochrome. The emitted signal (fluorescence) can be detected by conventional fluorimetric techniques. A typical set-up for an optical biosensor is shown in Figure 1. As depicted in this figure, the construction of the sensor for the fibre-optic transducer system is the main research and development component and the remaining instrumentation can be tailored depending on the required sensitivity and set-up.

Because of the availability of fluorescent-labeled antibodies and enzymes a number of research groups are actively engaged in the development of various optical biosensors using such probes (Arnold 1991, Hanbury *et al.* 1997, Grant and Glass 1999, Kumar *et al.* 2000). The main focus of these studies has been to design a suitable host matrix for entrapment of labeled antibodies and finally to couple the matrix on the surface of fibre optics for detection. Immobilization of labeled antibodies on the surface of optical fibre is desirable. The key focus has been the construction of sensors specific to a particular bioanalyte in a non-aqueous host matrix. Sol–gel matrices made from inorganic salts appear to be the most suitable host matrix and several groups have demonstrated their potential for the development of optical biosensors (Hanbury *et al.* 1997, Grant and Glass 1999, Kumar *et al.* 2000). In all these studies the reactivity of labeled antibody in such matrices was demonstrated to be similar to aqueous medium.

Ideally a biosensor should respond selectively, continuously, rapidly and without adding reagent. A reagentless biosensor would mean that the tip of the sensor, e.g., optical fibres could be placed in contact with sample inside the body (Coulet 1991) and this can also be realized in optical biosensors by the detection of fluorescence from biomolecules and drugs in *in-vivo* in tissue samples. In particular, *in-vivo* fluorescence spectroscopy of tissue due to the presence of intrinsic biomolecules for detection and characterisation of premalignant and malignant lesions is developing rapidly in the context of clinical diagnosis (Andersson-Engels and Wilson 1992). The intrinsic fluorescence arises due to the presence of biomolecules like proteins, enzymes, porphyrins, elastin and collagen, which are fluorescent in nature. The overall distribution of these biomolecules in different types of tissue will strongly influence fluorescence spectroscopy. This fluorescence is known as tissue autofluorescence and studies conducted by several groups have already demonstrated its usefulness towards detection of early stage cancer (Andersson-Engels *et al.* 1997, Papazoglou 1995, Andersson-Engels and Wilson 1992, Goujon *et al.* 2000). A typical tissue fluorescence detection set-up is shown in Figure 2. It is to be mentioned that this type of fibre-optic set-up would satisfy the requirement of direct, continuous, rapid and reagentless biosensors. This type of set-up can be classified as an extrinsic sensor. Owing to the availability of low-cost diode lasers,

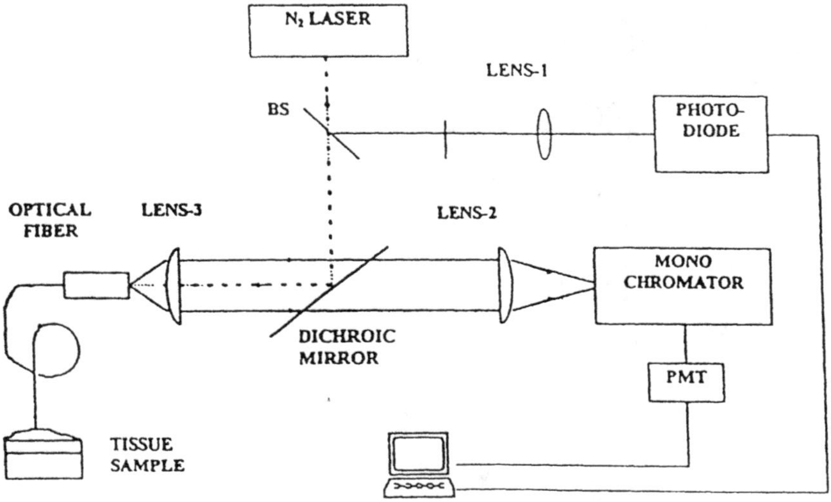

Figure 2. Schematic diagram of the experimental set-up for auto-fluorescence spectroscopy of tissues.

micro-optical components and very high sensitivity detectors, detection of fluorescence from such intrinsic fluorochromes has become feasible from tissues and various groups are conducting trials with prototypes in clinical laboratories in hospitals (Stummer and Baumgartner 2000, Leunig *et al.* 2000). An optical biosensor based on this arrangement can be useful for instant estimation of the concentration of photoactive-drug (photosensitizer) in photodynamic therapy of cancers.

In this chapter a brief overview on the current research activity based on two optical methods will be presented with the aim of widening the scope of optical biosensors. Since in the development of optical biosensors, design and construction of sensor material is a central problem, an attempt has been made to present certain basic aspects related to fluorescence molecules and their spectroscopic properties in order to support the design process.

1.1. Interaction of Light with Molecules in Relation to Fluorescence

When light interacts with atoms or molecules, the electric field of radiation tends to disturb or change the electron density of interacting atoms or molecules. A molecule has different electronic energy levels depending on its symmetry and upon interaction with electromagnetic field of light, either absorption or emission of radiation will occur and

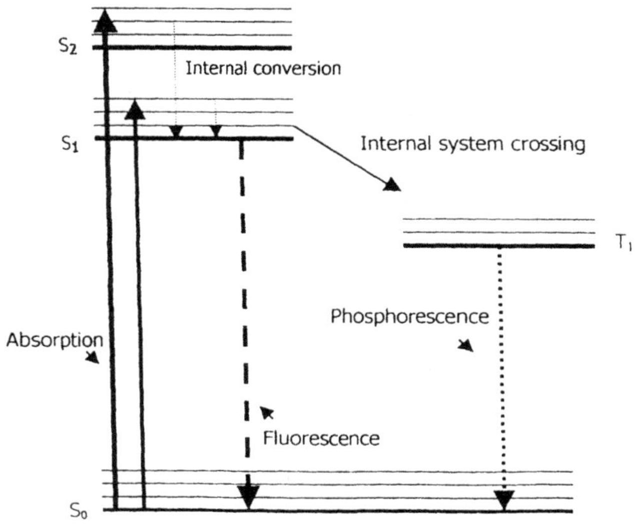

Figure 3. Basic processes involved in the interaction of a molecule with light.

the system can reach another stationary energy level. The interaction of light with a molecule leads to absorption only if a dipole moment is created and during the process of emission the dipole is destroyed. The different electronic levels comprise of various sublevels, viz. vibrational and rotational components. The amount of energy associated with each electronic level is of the order of 160–600 kJ/mole and the different sublevels viz., vibrational and rotational are 20–40 kJ/mole and about 0.08 kJ/mole respectively.

Figure 3 shows the different light scattering processes for fluorescence emission. When a molecule absorbs a light photon, it interacts with the dipole moment of the molecule and it undergoes various physical processes in different time scales ranging from sub-pico (10^{-12}) seconds to nano (10^{-9}) seconds. The first event is the instantaneous absorption of light energy (10^{-15} s) by the interacting molecule in which the molecule is excited to the first singlet electronic state (S_1) from the ground electronic state (S_0). In the excited state the molecule remains for a definite but short time period ranging from picoseconds to few nanoseconds and then deactivates to the ground state (S_0) either by radiative or non-radiative processes. The radiative decay process is known as fluorescence that can be detected and analysed by a combination of a spectrometer (analyser) and a photomultiplier (detector) system. The intensity of spectrally allowed transitions is dictated by the molecular symmetry of the molecule under investigation. Emission data when derived as a function of scanning wavelength generate an emission spectrum which is essentially a plot of the fluorescence intensity versus scanning wavelength. The spectral positions of emission bands are characteristic of the molecule and also very sensitive to the microenvironment. If the molecule undergoes intersystem crossing from singlet excited state (1S) to triplet excited state (T), then because of the relatively high lifetime of this state it can either deactivate by phosphorescence or degradation.

1.2. Intensity of Fluorescence Signals

Fluorescence intensity is quantitatively dependent on the Lambert–Beer law and quantum yield. In dilute solutions or homogeneous suspension intensity is directly proportional to its concentration. Since the observed relative fluorescence intensity is also dependent on the type of light source, optical components (i.e., focussing, collection optics, monochromators) and detection systems, fluorescent reference standards are therefore required for quantitation of fluorescence signals of unknown fluorochromes or even known fluorochromes in unknown environment. The variation of dye fluorescence with environment and effects of quenching processes results in difficulty in quantitation. Because of this, fluorescent dyes are usually modified for labeling of a particular biomolecule in such a way that the fluorescence property remains almost independent of the environment. The fluorescence signal is usually very high as compared to absorption, but often due to the presence of a relatively strong background signal, either due to scattering of the incident light from other molecules or from intrinsic fluorophores, requires judicious subtraction of the background intensity. This can be achieved by using filters, a monochromator and choosing appropriate combination of fluorophores and excitation wavelength. The contribution due to the latter effect can be achieved by selecting dyes that have a larger Stokes shift, i.e. the difference between the excitation and emission wavelengths. For the detection of analytes in the low concentration range using extrinsic fluorescent molecules, the quantum yield of fluorescence is also expected to be very high.

1.3. Photobleaching

Due to continuous illumination, a fluorophore may suffer irreversible destruction and therefore the fluorophore under investigation may become a limiting factor. Usually photo-bleaching originates from the relatively long-lived (micro- to milliseconds) triplet excited state (Figure 3) and therefore, those fluorophores where intersystem crossing efficiency is minimum would be suitable for practical optical biosensors. In general, photo-bleaching can be avoided by lowering the intensity of excitation and enhancing the detection efficiency. Usually this is achieved by replacing commonly used photomultiplier systems with ultra-sensitive detectors like multistage cooled photomultipliers and CCD cameras, in combination with high-quality optical components. However, the concentration of fluorophore should not be increased as it would cause marked changes in the chemical and optical characteristics.

These fluorescence spectral properties viz., spectral position, intensity, quenching and photobleaching are very important parameters which need to be considered for proper selection of a molecule as a probe in the development of biosensors.

2. FLUORESCENCE MOLECULES

A variety of fluorescence molecules are available for studies at a molecular level using fluorescence-based techniques. For example, fluorescent dyes are being used extensively for studying protein structure (Haugland 1982, Lakowicz 1999). In immunofluorescence studies fluorescent-labeled antibodies are available, in addition fluorescent probes for enzymes, proteins, DNA hybridization, growth factor, drugs peptides, lipids, hormones, etc., are also available. Moreover, imaging of intracellular fluorochromes with the finest detail is also illustrated in the literature. Therefore, depending upon the origin of these molecules, fluorochromes are classified either as endogenous (intrinsic) or exogenous (extrinsic). It is necessary to understand their properties so that the appropriate selection of a fluorescent probe is made.

2.1. Endogenous (Intrinsic) Fluorochromes

Intrinsic fluorochromes are constitutive molecules of cells and tissues. Tissues are cooperative assemblies of different cells of multicellular organisms and as such exist in various compositions. The cells are held together by extracellular macromolecules in the matrix. The extracellular matrix primarily consists of fibrous proteins viz., collagen, elastin,

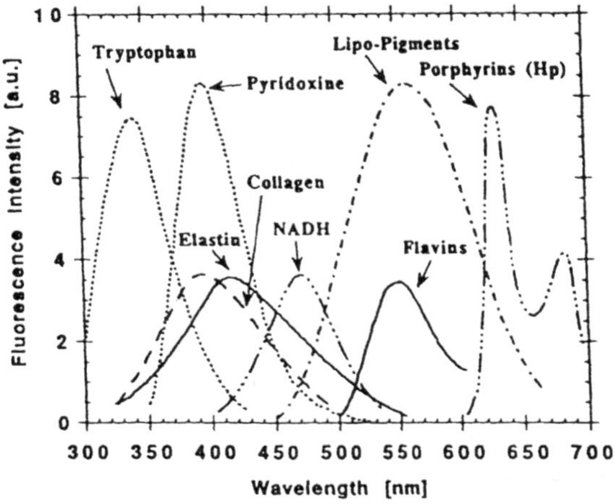

Figure 4. Fluorescence spectra of important biomolecules (Wagnières *et al.* 1998).

fibronectin, etc. The other major constituents of the extracellular matrix are polysaccharides and glycosaminoglycans. The structural proteins, collagen and elastin can undergo both quantitative and qualitative change under pathological conditions. For example, normal aortic wall is composed of up to 20% elastin and 30% of collagen by dry weight, whereas in atherosclerotic lesions collagen can account for up to 60% of the dry weight with less than 10% of elastin. The matrix containing extracellular constituents viz., collagen, elastin plays an important and complex role in regulating the behaviour of the cells, influencing their development, migration, proliferation and metabolic functions. In view of this, information about both the cellular as well as the extracellular matrix is equally important in understanding the biochemical processes which precede and accompany the onset of disease (Manoharan *et al.* 1996). In addition, cells contain a number of structural and functional macromolecules like proteins, enzymes, the reduced form of nicotinamide adenine dinucleotide (NADH), flavins and heme proteins, porphyrins, etc. Figure 4 shows the distinct fluorescent emission spectra of these biomolecules.

Several constituents of proteins viz., amino acids like tryptophan, tyrosine and phenylalanine are fluorescent in nature. The fluorescence behaviour of tryptophan, one of the important fluorochromes in protein, has been extensively investigated in order to understand the structure–function relationship of protein (for a review, see Lakowicz 1999). Therefore, in principle, fluorescence originating from a specific protein

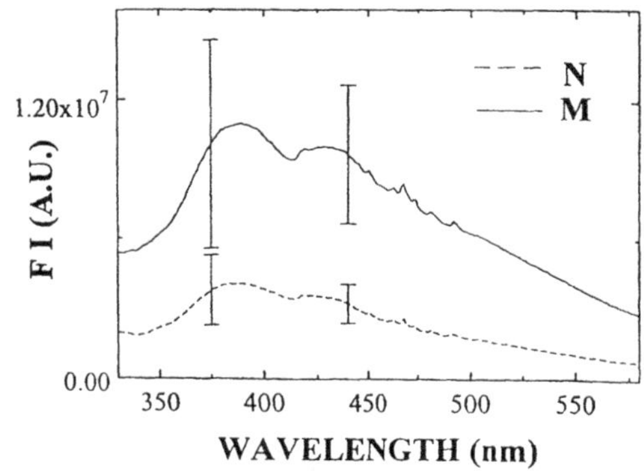

Figure 5. Mean fluorescence spectrum of normal and malignant tissues with 310 nm excitation. The malignant and the normal tissue spectra are the average of the spectra of tissue samples from 11 patients (Majumder *et al.* 1999).

can be utilized for optical biosensing. Tissues, depending upon their origin, contain various types of proteins and other biomolecules. Attempts are being made to use tissue autofluorescence spectroscopy for direct detection and characterisation of early stage cancers. This is because the characteristic fluorescence from tissue will depend on the relative concentration, spatial distribution of biomolecules and metabolic status. Tissue fluorescence has been used to differentiate normal and abnormal tissue in human breast, lung (Tang *et al.* 1989, Lam *et al.* 1993), bronchus, oral mucosa (Ingrams *et al.* 1997) and gastrointestinal tract (Schomacker *et al.* 1992, Panjehpour *et al.* 1996, Romer *et al.* 1995). The initial results of fluorescence spectroscopy have been shown to be promising for clinical diagnosis of pre cancer and cancer of these organs. (Onizawa *et al.* 1996, Ramanujan *et al.* 1996, Braichotte *et al.* 1991, Chaudhury *et al.* 2001). Figure 5 shows the 310 nm excited mean fluorescence emission spectra of the malignant and normal tissue samples (Majumder *et al.* 1999). Autofluorescence from malignant and normal human breast tissues excited in the UV region at 310 nm showed significant difference viz., the malignant tissues was reported to be more fluorescent compared to the normal tissues. The individual spectrum depicted peaks at 390 nm and 440 nm and these peaks were assigned to structural proteins and co-enzyme (NADH). Since instrumentation and the fibre-optic arrangement are almost similar the set-up can be used for development of an optical biosensor.

NADH has a relatively weak fluorescence but detection of sub-micromolar concentrations can be achieved. The intracellular levels of NADH have been measured by fluorescence spectroscopy (Scheper *et al.* 1989). Arnold (1991) has made an attempt to develop a biosensor by immobilizing the enzyme lactate dehydrogenase (LDH) at the distal end of optical fibre. LDH catalyzes the conversion of lactate to pyruvate with the cofactor NAD^+ which converts to NADH. Higher intensity is expected for higher concentration of lactate. Although this phenomenon has been known for a long time, due to certain intrinsic limitations NADH fluorescence-based biosensors could not make significant progress until now.

2.2. Fluorescence due to Pro Drug 5-Aminolevulinic Acid (5-ALA)

5-Aminolevulinic acid (5-ALA) is a precursor in biosynthesis of heme. External administration of 5-ALA, a prodrug which can regulate the biosynthesis of heme in cells, is expected to result in different rates of synthesis in normal and in cancerous cells (Peng *et al.* 1997). Administration of ALA actually gradually accelerates the synthesis of a fluorophore protoporphyrin-IX (PP-IX) and inhibits its ultimate conversion to iron porphyrin (heme). In principle, the accumulation of PP-IX can be further enhanced by modulation of the biosynthetic pathway by addition of iron chelators such as EDTA. The characteristic fluorescence from PP-IX, i.e. red emission around 635 nm, can be useful in differentiating normal and abnormal tissue. This is because the concentration of ALA-induced fluorophore, i.e. PP-IX, can be markedly different depending upon the physiology and pharmacokinetics that vary in spatial distribution and with nature of disease in tissues (Hayata *et al.* 1996). Due to pioneering work of Dougherty *et al.*, at Buffalo, USA since 1970, photodynamic therapy (PDT) has now reached a level of clinical practice in several countries (Dougherty *et al.* 1978). Photofrin II, a porphyrin-based fluorescent molecule for photodynamic therapy (PDT) is being used as photosensitizing fluorescent drug in more than ten countries for certain types of cancers. The prodrug 5-ALA is being extensively investigated for detection of early stage cancerous lesions in different organs (Heil *et al.* 1997). Photodynamic therapy is a new treatment modality and requires administration of a photosensitizing drug which is fluorescent in nature. After an interval of 24–48 h, the target volume is exposed to light of suitable wavelength, preferably from a

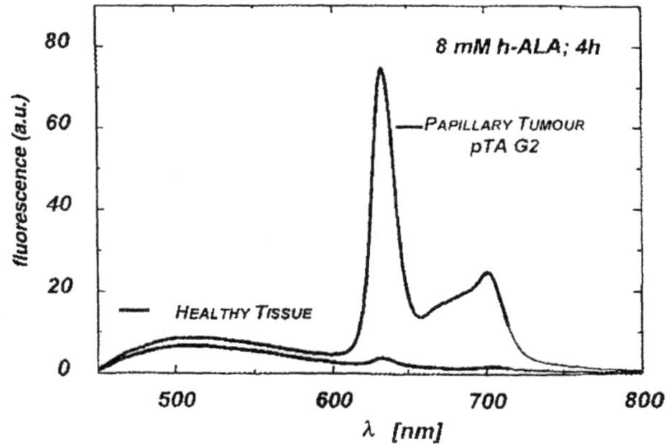

Figure 6. The fluorescence spectrum of a small papillary tumour (pTa G2, top line), compared to the surrounding healthy tissue (bottom line) measured *in vivo* after 4 hours of instillation of 8 mM of h-ALA, a more effective prodrug than 5-ALA. The absence of PP IX fluorescence in the healthy tissue indicates the high selectivity of h-ALA (Van den Bergh *et al.* 1999).

laser coupled to optical fibre. Due to the ease of handling of patients for photodynamic treatment, this prog-drug can be used for detection, characterisation of lesions and even for treatment planning. Results of clinical trials conducted at various centres have already started appearing in the literature (Schuimaker *et al.* 1996, Peng *et al.* 1997, Leunig *et al.* 2000). Figure 6 shows the differential effect of ALA hexyl-ester (h-ALA), a derivative of ALA, on fluorescence from normal tissue and a papillary tumour (Van den Bergh *et al.* 1999) Characteristic fluorescence emission bands around 635 and 700 nm due to accumulation of PP-IX from tumour volume are observed. Cancer screening and early detection will have major importance in the survival of patients with cancer. Thus fluorescence detection methods are expected to have great implication in management of cancer.

2.3. Extrinsic Fluorophores

The selectivity of antibodies towards particular protein analytes is widely used for construction of optical biosensors (Arnold 1991). In proteins, although there are several important reactive sites, most of the reaction sites occur at either amino groups of lysine or thiol groups of cysteine. Therefore depending on the availability of these reaction sites in proteins either amines or thiols directed fluorochromes are chosen for the purpose of conjugation with antibodies. The conjugated probe is expected to have high yield of fluorescence and reactivity with enzymes and proteins. Amine-reactive fluorescence probes are widely used to prepare conjugates with proteins, peptides, ligands and oligonucleotides. These are mostly used in immunochemistry, fluorescence *in situ* hybridization, cell cycle analysis, etc. In general, the fluorophores are modified to introduce a functional group to enable the molecule to undergo conjugation at the target sites in biomolecules. These fluorescence-labeled conjugates are expected to be stored and experimented in extreme conditions and therefore, their stability is particularly important.

Different methods are being experimented with to systematically monitor the changes occurring in fluorescent-labeled compounds upon binding with the target analyte. Fluorescence energy transfer (FRET) is one important phenomenon in which the fluorescence energy from the labeled antibody (donor) is transferred to the analyte (acceptor) (Haugland 1982). The donor and acceptor must be in close proximity. Excitation of the donor molecule will thus produce emission from the acceptor that ordinarily would not occur in the absence of FRET. The decrease in fluorescence intensity from the donor can be measured and correlated with the concentration of analyte in the medium. Fluorescence quenching is another method in which the fluorescence of the labeled complex is decreased by the presence of another molecule or analyte in an optical transducer system.

2.4. Selected Examples of Potential Extrinsic Fluorophores

2.4.1. Fluorescein

This dye is the most common fluorescence agent which has been used extensively as molecular probe for proteins. This is because of its high absorption coefficient and good water solubility. In particular, the quantum yield of fluorescence in solution is very high (Table 1). The absorption and emission maximum of fluorescein are at 494 nm and 520 nm, respectively. The quantum yield of fluorescence, although quenched more than 50% on conjugation with protein, is sufficiently high for all routine investigations. Fluorescein dye has some drawbacks: the phenolic group has a pKa of about 6.4 and it has been reported that the absorbance and fluorescence both are lowered at pH below 7. In addition, this compound undergoes rapid photo bleaching.

Table 1. Physicochemical properties of selected fluorescent molecules[a]

Fluorescence molecule	Mol. wt	Abs_{max} (nm)	$Fluo_{max}$ (nm)	Solvent	Extinction coeff. ($\times 10^3$/M cm)
Fluorescein	332	494	514	DMF	80
Fluorescein-isothiocyanate	389	494	519	DMF	73
Fluoresceine carboxysuccinimidyl ester	473	495	519	DMF, DMSO	74
Rhodamine green carboxylic acid	411	502	524	DMF	74
Texas red sulfonylchloride	625	587	602	$CHCl_3$	90
Texas red succinimidyl ester	817	583	603	MeOH	116
Cascade blue acetyl azide	607	391	410	H_2O, MeOH	29

[a] Parameters are taken from Molecular Probes Catalog (1996), in the 6th edition of the *Handbook of Fluorescent Probes and Research Chemicals* edited by R.P. Haugland. The extinction coefficient determines the extent of light absorption by the molecule.

2.4.2. Fluorescein Isothiocyanate (FITC)

Fluorescein isothiocyanate (FITC) is one of the most useful fluorescent labeling reagents. This derivative of fluorescein is extensively used to prepare antibody conjugates for use in immunofluorescence assays. Synthesis of FITC often leads to formation of two isomers which are spectrally indistinguishable. Out of the two isomers of FITC, 5-isomer is commonly used because it is easier to isolate. Various other types of substituents, e.g., succinimidyl esters, chlorotriazines in FITC are also used for labeling of proteins and are relatively stable (Blakeslee and Baines 1976). Many commercial preparations of fluorescein bioconjugates are available (Molecular Probes Catalog 1996, Calbiochem 1997).

2.4.3. Rhodamine Green and its Derivatives

Rhodamine green is another potential fluorophore which is completely insensitive to pH ranging from 4 to 9. The bioconjugate of rhodamine is relatively more stable. The absorption and fluorescence emission spectral maxima are slightly red-shifted by 7 nm as compared to fluorescein. The fluorescence of conjugates with protein tend to quench and these conjugates often precipitate in solution. Tetramethylrhodamine (TMR) is an important fluorophore especially in applications involving labeled proteins, antibody and avidin in immunochemistry. The fluorescence quantum yield of TMR conjugates is one fourth of that of fluorescein conjugates. But TMR conjugates can be easily excited by using a simple mercury arc lamp and is more photostable than fluorescein. Similar to FITC, tetramethyl rhodamine isothiocyanate (TRITC) also exists in two isomers. The absorption spectrum of TRITC conjugate is complex and often splits into two bands at 520 and 550 nm depending on the environment. TRITC conjugates can also be excited by a mercury arc lamp and despite their low quantum yield, strong fluorescence intensity can be easily detected. The typical extinction coefficient of TRITC conjugates is about 80 000 M^{-1} cm^{-1}. It is to be mentioned that the higher the extinction coefficient, the higher is the ability of a molecule to absorb light. This means even an ordinary light source such as a halogen lamp can be used for excitation.

2.4.4. Texas Red and its Derivatives

Texas red is derived from sulforhodamine 101 and emits at a longer wavelength (> 500 nm), and the fluorescence yield of its protein conjugates is usually higher than that of TRITC. Texas red conjugates are particularly well suited for the 568 nm emission line of Ar–Kr laser. Texas red sulfonyl chloride is a isomeric sulfonyl chloride of sulforhodamine 101. Although the aqueous solution of this compound is quite unstable in water, its protein conjugates are very stable and even survive complete protein hydrolysis. Texas red-x-succinimidyl ester, another derivative of texas red, shows higher fluorescence yield than texas red.

2.4.5. Cascade Blue

Cascade blue fluorophore emits in the blue region (410 nm), it has a high absorption coefficient and unlike many other dyes does not quench fluorescence of conjugates. The protein conjugate of this fluorescence molecule can be excited with a low-power Hg arc lamp. Cascade blue acetyl azide may be useful for identifying proteins located on extracellular cell surfaces. This derivative shows significantly more fluorescence even at low degrees of substitution as compared to other fluorochromes. The extinction coefficient and quantum yield of cascade blue is about 28 000 M^{-1} cm^{-1} and 0.54, respectively.

Many of these fluorescence molecules and their derivatives coupled to various antibodies are commercially available (Sigma Chemicals 2002, Molecular Probes Catalog 1996). Although many of these fluorescent dyes are quite suitable for studies in the field of molecular biology, due to several limitations the demand for new dyes with better sensitivity and specificity is ever-increasing. For the development of biosensors, consideration of stability and sensitivity is crucial because accurate detection of low concentrations of analyte will depend on these aspects. The desirable properties of an ideal fluorescent dye specific to particular application in optical biosensor are (i) high quantum yield (ii) minimum quenching on conjugation (iii) minimum photobleaching (iv) higher Stokes' shift and (v) high solubility of the conjugates. Some of the important properties of selected fluorescent molecules are shown in Table 1 (above).

3. CURRENT STATUS OF THE DEVELOPMENT OF OPTICAL BIOSENSORS

Usually the sensor material containing fluorescent-labeled protein or antibodies need to be immobilized at the distal end of the optical fibre. The labeled conjugate is expected to selectively allow absorption of the analyte and the subsequent changes in optical properties need to be monitored for quantitation. This approach is under intensive investigation in various research centres and in the following sections few selected examples are briefly mentioned.

3.1. Sol–gel Matrices for Fibre-optic Detection

A simple, stable host which can be immobilized at the surface of optical fibre is ideally suitable for the transducer interface in optical biosensing of analytes. Thus fluorescent-labeled antibody immobilization technology is one of the key aspects in biosensor development. It has been emphatically demonstrated that the sol–gel matrix is a suitable host for the immobilization of fluorescent-labeled macromolecules (Ellerby *et al.* 1992, Wang *et al.* 1993). The sol–gel matrix is made of an inorganic oxide of glasses viz., tetramethyl orthosilicate (TMOS). This type of glass can be conveniently fabricated at room temperature. Sol–gel glasses are highly porous structures, and high-molecular-weight dopant molecules can be trapped inside these pores, if added during the preparation stage. The sol–gel glasses are very stable and retain the reactivity of biomolecules (Dave *et al.* 1996). A variety of different proteins can be immobilized in sol–gels and thereby preventing the self-aggregation effect. These entrapped proteins, enzymes and other molecules are functionally active within these glasses.

3.2. Encapsulation of Biomolecules and Sol–Gel Biosensors

A variety of biomolecules can be trapped in the sol–gel matrix (Avnir *et al.* 1994). The biomolecules added to the sol–gel are trapped within the growing polymeric network as the final porous gel forms. The dopant molecule, e.g. protein, is able to bind low-molecular-weight ligands as well as carry out catalysis of substrates in a way exactly similar to aqueous solutions. The molecular sizes of the substitutes and ligands are considerably smaller than the average pore diameters and they are mobile in the solvent-filled pore. The inorganic matrix makes the resulting solutions chemically robust and thermally insensitive.

Since the pore diameter is less than wavelength of light, the sol–gel glasses are optically transparent. This enables one to monitor spectroscopically (by absorption and fluorescence) the nature of molecules entrapped in a sol–gel prepared from TMOS. The sol–gel glass can also form a variety of shapes and forms, e.g., films and monoliths. Myoglobin (Mb) was among the first proteins to be entrapped and it was shown that the absorption spectral characteristics of Mb remained unchanged (Dave *et al.* 1998). It was shown that the solution chemistry of myoglobin, for example, the chemical redox reaction with dissolved oxygen, can still be observed in a sol–gel matrix as depicted in Figure 7. The results clearly show the characteristic oxidized/reduced states of Mb. By monitoring the changes in absorbance at 436 nm due to deoxymyoglobin, the unknown concentration of dissolved oxygen can be determined. This redox state is highly reversible, i.e. reversibility of oxidized to reduced state, and has been demonstrated by chemical methods in such a matrix environment.

This biosensor system would require a spectrophotometric arrangement for detection and quantitation. This technique would be less costly than a fluorescence detection system. Such a biosensor for oxygen detection could be very useful for the detection of molecular oxygen at a cellular level, particularly where therapeutic efficacy is dependent

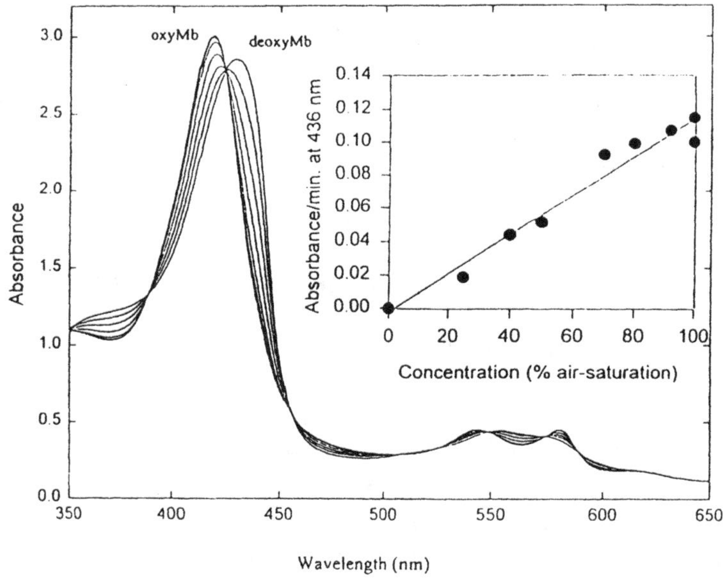

Figure 7. Absorption spectral changes in deoxy Mb in presence of dissolved oxygen in sol–gel. The inset shows the calibration curve drawn with OD. At 436 nm (Dave *et al.* 1998).

upon the availability of oxygen. For example, in photodynamic therapy, pre-administered photoactive drug (photosensitizer), light and molecular oxygen act together within the tumor volume to produce singlet oxygen, a reactive species of oxygen that is toxic to cells.

Cytochrome C has also been reported to be trapped in sol–gel matrix in the form of a thin film (Dave *et al.* 1998). The absorption spectrum for a reference solution of the protein dissolved in buffer has a Soret maximum at 407 nm, and the absorption maximum of the protein in thin film is at 405 nm (Figure 8). It has been further shown that air oxidation of the reduced form and the original oxidized form are reversible in both solution and thin film. Interestingly, the oxidation reduction reaction is significantly faster in thin film than under solution conditions.

Hanbury *et al.* in 1997 developed a self-contained fibre-optic immunosensor to measure 16 500 Da protein myoglobin. The protein molecule was detected by using cascade-blue-labeled antibody within polyacrylamide gel adsorbed at the tip of optical fibre. This polyacrylamide gel is similar to a sol–gel and enables spectroscopic monitoring of entrapped molecules. The analytical signal was derived from fluorescence

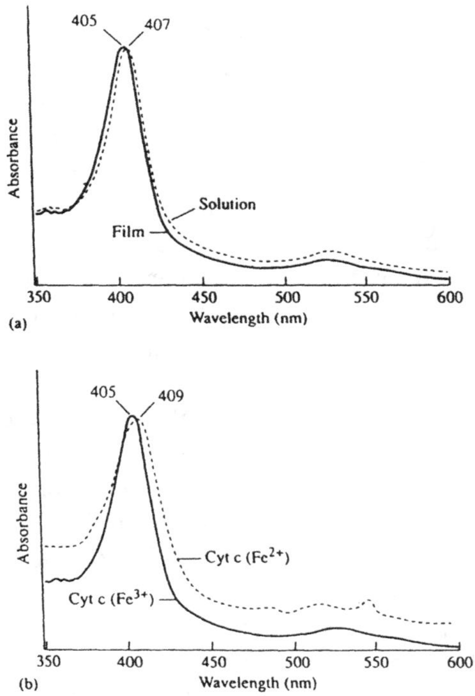

Figure 8. Absorption spectra of cyt c immobilized in sol–gel films: (a) comparison of spectrum in solutions and film; (b) reversibility of redox states of cyt c with sodium dithionite (Dave *et al.* 1998).

energy transfer between the fluorophore cascade blue and the heme group of myoglobin. Fluorescence intensity was quenched depending on the antibody labeling condition and amount of antibody incorporated. The lowest concentration of myoglobin detected was 5 nmol/L. The fibre-optic myoglobin sensor demonstrated the feasibility of measuring a large molecular mass protein analyte. The fibre-optic sensor was found to be stable in buffer and delivered a concentration-dependent fluorescence response to myoglobin. Further research work is necessary to test its suitability for whole blood measurements.

In thrombolytic therapy for the treatment of stroke, the thrombolytic agent cleaves the circulating plasma proenzyme plasminogen, but often binds with fibrinogen resulting in its degradation. One of the degraded products is the D-dimer fragment which in turn may lead to undesirable hemorrhage. A fibre-optic biosensor for the detection of fibrinolytic products produced during lysis of soft blood clots has been developed

(Grant and Glass 1999). The biosensor was constructed particularly for D-dimer antigens, which form from the dissolution of cross-linked fibrin clots. The presence of D-dimer antigens above a threshold level is a clinical diagnostic tool used to determine the presence of such occlusions following a stroke. Fluorescein-labeled D-dimer antibodies are trapped inside the sol–gel and immobilized on the tip of an optical fibre. The decrease of fluorescein intensity with the increase of dimer antigens was correlated with the concentration of D-dimer. Sensitivity of D-dimer was in the clinically relevant range (0.5 μg/ml to higher levels) and linearity has been demonstrated in phosphate buffer, human plasma and whole blood. The D-dimer antibodies remained partially viable at least four weeks while encapsulated in the sol–gel network when stored at 4°C in PBS solution and the response time of the set-up was 30 s. This experimental set-up could be used for diagnosis of arteriovenous clot types and for monitoring thrombolysis in stroke patients undergoing thrombolytic therapy. This sensor could be coupled along with the catheter during treatment of stroke.

Kumar *et al.* (2000) have reported successful immobilization of cholesterol oxidase and horseradish peroxidase in thin films made from tetraethyl orthosilicate (TEOS). The activity of enzymes did not undergo any change in the non-aqueous medium. The cholesterol activity has been assayed as a function of concentration of cholesterol using spectrophotometric method. The rate of the reactions were compared with electrochemical technique and demonstrated linearity within 2–10 mM. Most importantly, all these films retained the activity of enzymes up to four months at room temperature.

4. PROSPECTS OF OPTICAL BIOSENSORS IN QUANTITATION OF PHOTOACTIVE DRUGS

During the last decade, photodynamic therapy of cancers has made significant progress (Schuimaker *et al.* 1996). However, this treatment protocol is relatively rather simple but still very much empirical in nature. The therapeutic efficacy can be improved significantly if direct estimation of the concentration of photomedicine in cells/tissues is made before treatment. These photoactive drugs are generally porphyrin- and phthalocyanine-based compounds, e.g., hematoporphyrin derivative,

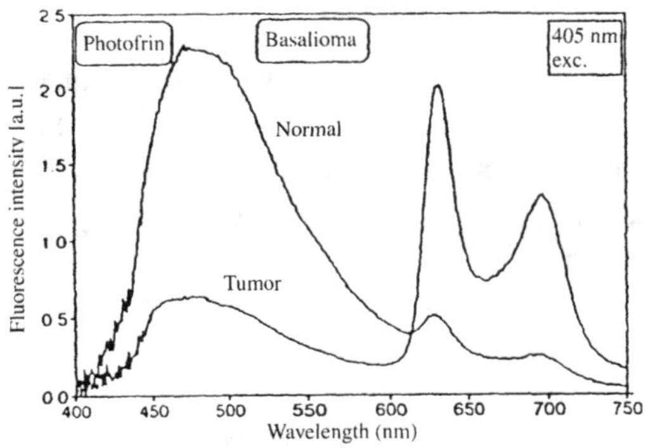

Figure 9. Fluorescence spectra of superficial basal cell carcinoma after 48 h of administration of photofrin II (Andersson *et al.* 1992).

mesotetrahydroxyphenyl chlorin, tin etioporphyrin, zinc tetraphenyl porphyrin and aluminium phthalocyanine. These photomedicines have strong absorption and fluorescent characteristics in the UV–visible range which can be detected directly by placing optical fibre. Estimation of oxygen in photodynamic therapy is another important parameter which strongly influences the efficacy of therapy. This can also be achieved by such fibre-optic devices. A fibre-optic sensor can be directly implanted at the treatment site to obtain the information for estimation of the light dose to be delivered. In addition, the progress of treatment can also be monitored by quantitating the characteristic fluorescence signal of drug as the uptake and retention in surrounding normal tissues are generally low resulting in the decrease of the fluorescence signal. Figure 9 shows the fluorescence spectra of a drug from normal and tumor sites. These spectra clearly show emission bands of hematoporphyrin derivative from tumour tissues whereas the autofluorescent band appeared only in normal tissue. With progress of the treatment the characteristic fluorescence intensity will decrease and eventually the spectral features will become almost the same as normal tissues where the fluorescence is basically due to the tissue autofluorescence.

5. TRENDS AND PROSPECTS

This chapter has reviewed primarily the current status of development of optical biosensors based on the retention of biochemical activity of fluorescent-labeled antibodies in inorganic materials, viz. sol–gel matrices as well as in polyacrylamide gel. The experimental data demonstrate that various types of fluorescent-labeled proteins and antibodies can be trapped within the sol–gel matrix. The proteins are remarkably stable, undergo characteristic reversible reactions similar to that in aqueous solutions and most importantly their spectroscopic properties are not affected. Results of these initial experimental studies are very promising and it appears that optical biosensors for clinical use could be realized in the near future.

There are few other areas where optical biosensors can have immediate applications in detection and management of cancers. Biosensors for oxygen measurement can be particularly important for treatment planning in photodynamic therapy. The accurate estimation of oxygen can be achieved by using a myoglobin biosensor. Fibre-optic probes can be useful for the direct determination of fluorescent molecule (drug) in target tissues (tumor). The presence of oxygen is particularly important because the photon energy absorbed by the drug (photosensitizer) is to be ultimately transferred to molecular oxygen for generation of singlet oxygen, the cytotoxic species. Estimation of both the parameters are crucial for treatment planning, i.e., time of treatment after administration of photoactive drug and light dose (exposure time), and at present no method is available and the treatment procedure is just empirical in nature.

Optical biosensors can also help in the detection of early stage cancer. This is because with the increasing understanding of processes related to formation of cancer, its proliferation and its regulation, several tumor suppressor genes have been identified by researchers. p53 is one such important gene which has been shown to be affected in a variety of cancers viz., colon, breast, lung, bladder, brain, pancreas, stomach and many others (Vogelstein *et al.* 2000). FITC-labeled antibodies for these oncogene and related proteins are commercially available for studies in molecular biology. Optical biosensors for detection of cancer cells based on such markers can be created using immobilization techniques. Monitoring the level of p53 can also help in evaluating the progress of treatment.

Certain practical aspects of optical fibres are also worth considering right from the beginning of sensor construction. This is because of the high price and fragile nature of these sensors, which make them unsuitable in some cases. However, depending on the spectral nature of fluorescent molecules, glass fibre or even plastic optical fibre can be used. Low-cost tungsten–halogen lamps can be used if measurements are to be carried out based on absorption, otherwise low-pressure mercury arc lamps or low-cost laser are required for fluorescence detection. Most importantly, the immobilization method might differ depending on these optical fibres.

Fluorescent molecules and the development of suitable methods for immobilization and detection of specific bioanalytes still remain the centre of research activities in optical biosensor. However, the complete development of a biosensor device for clinical use will require effective interfacing physics, chemistry and biology with electronics.

ACKNOWLEDGEMENT

The author is thankful to Dr. T.L. Mathew, Director, Institute of Nuclear Medicine and Allied Sciences, Delhi for his constant encouragement and research interest in the area of biosensors and bioelectronics.

REFERENCES

Andersson, T., R. Berg, J. Johansson, D. Killander, K. Svanberg, S. Svanberg and Y. Yuanlong (1992). Photodynamic therapy in interplay with fluorescence diagnosis in the treatment of human superfical malignancies. In *Optical Methods for Tumor Treatment and Detection*, *SPIE*, **1645**, 187–198.

Andersson-Engels, S., and B.C. Wilson (1992). *In vivo* fluorescence in clinical oncology: fundamental and practical issues. *J. Cell Pharmacol.*, **3**, 66–79.

Andersson-Engels, S., C. Klineteberg, K. Svanberg and S. Svanberg (1997). In vivo fluorescence imaging for tissue diagnosis. *Phys. Med. Biol.*, **42**, 815–824.

Arnold, M.A. (1991). Fluorophore- and chromophore based fiber optic biosensor. In *Biosensor principles and application*. L.J. Blum and P.R. Coulet (Eds.), Marcel Dekker, New York, pp. 195–212.

Avnir, D., S. Braun, O. Lev and M. Ottolenghi (1994). Enzymes and proteins entrapped in sol gel materials. *Chem. Mater.*, **6**, 1605–1614.

Blakeslee, D., and M.G. Baines (1976). Immunofluorescence using dichlotriziny1 amino fluorescence (DTAF). Preparation and fractionation of labeled IgG. *J. Immuno. Meth.*, **13**, 305–320.

Braichotte, D., G. Wagneeres, P. Monnier, M. Savary, R. Bays, H.E. van den Bergh and A. Chatelian (1991). Endoscopic tissue autofluorescence measurements in the upper aerodigestive tract and the bronchi. *Proc. SPIE*, **1525**, 211–218.

Calbiochem (1997). Oncogene research products, PO Box 12087, La Jolla, CA 92309–2087, USA.

Chaudhury, N.K., S. Chandra and T.L. Mathew (2001). Oncological applications of biophotonics: Problems and prospects. *Appl. Biochem. Biotech* **96**, 183–204.

Coulet, P.R. (1991). What is a Biosensor?. In *Biosensor principles and applications*. L.J. Blum and P.R. Coulet (Eds.), Marcel Dekker, New York, ch. 1, pp. 1–6.

Dave, B.C., B. Dunn, J.S. Valentine and J.I. Zink (1996). In *Nanotechnology: Molecularly designed materials*. G.M. Chow and K.E. Gonsalves (Eds.), American Chemical Society, Washington, DC, p. 351.

Dave, B.C., B. Dunn, J.S. Valentine and J.I. Zink (1998). Sol–gel matrices for protein entrapment. In *Immobilized Biomolecules in Analysis – A practical approach*. T. Cass and F.S. Ligler (Eds.), Oxford University Press, New York, pp. 113–134.

Dougherty, T.J., J. Kaufman, A. Goldfrab, R. Weishaupt, D.G. Boyle and A. Mittelman (1978). Photoradiation therapy for the treatment of malignant tumours. *Cancer Res.*, **38**, 2628–2635.

Ellerby, L.M., C.R. Nishida, F. Nishida, S. Yamanbaka, B. Dunn, J.S. Valentine and J.I. Zink (1992). Encapsulation of proteins in transparent porous silicate glasses prepared by the sol gel method. *Science*, **255**, 1113–1115.

Goujon, D., M. Zellweger, H. van den Bergh and G.A. Wagnières (2000). Autofluorescence imaging in the tracheobronchial tree. *Photodynamic News*, **3**(3), 11–14.

Grant, S.A., and R.S. Glass (1999). Sol–Gel based biosensor for use in stroke treatment. *IEEE Trans. Biomed. Eng.*, **46**, 1207–1211.

Guilbault, G.G. (1972). Analytical uses of immobilized enzymes. In *Enzyme Engineering*. L.B. Wingard Jr. (Ed.), Interscience, New York, pp. 361–376.

Hanbury, C.M., W.G. Miller and R.B. Harris (1997). Fiber-optic immunosensor for measurement of myoglobin. *Anal. Chem.*, **43**, 2128–2136.

Haugland, R.P. (1982). Covalent fluorescent probes. In *Fluorescence Proteins and Nucleic Acids*. R.F. Steiner (Ed.), Plenum Press, New York, pp. 29–58.

Hayata, Y., H. Kato, K. Fursue, Y. Kusunoki, S. Suzuki and S. Mimura (1996). Photodynamic therapy of 168 early stage cancers of the lung and oesophagus: a Japanese multi-center study. *Lasers Med. Sci.*, **11**, 255–259.

Heil, P., S. Stocker, R. Sroka and R. Baumgartner (1997). In vivo fluorescence kinetics of porphyrins following intravesical instillation of 5-aminolevulinic acid in normal and tumour bearing rat bladders. *J. Photochem. Photobiol. B* **388**, 158–163.

Ingrams, D.R., J.K. Dhingra, K. Roy, D.F. Perrault, I.D. Bottrill, S. Kabani, M.M. Pankratov, S.M. Shapshay, R. Manoharan, I. Itzkan and M.S. Feld (1997). Autofluorescence characteristics of oral mucosa. *Head Neck J. Sci. Specialities*, **19**, 27–32.

Johnson, P.A., T.E. Barber, B.W. Smith and J.D. Wineford (1989). Ultralow detection limits for an organic dye determined by fluorescence spectroscopy with laser diode excitation. *Anal. Chem.*, 61, 861–863.

Kumar, A., R. Malhotra, B.D. Malhotra and S.K. Grover (2000). Co-immobilisation of cholesterol oxidase and horse radish peroxidase in a sol gel film. *Anal. Chem. Acta*, **414**, 43–50.

Lakowicz, J.R. (1999). *Principles of Fluorescence Spectroscopy*, 2nd edition, Plenum Press, New York.

Lam, S., C. Mac Auley and B. Paleic (1993). Detection and localization of early lung cancer by imaging technique. *Chest*, **103**, 12S–14S.

Leunig, A., C.S. Betz, M. Mehlmann, H. Stepp, G. Grevers and R. Baumgartner (2000). Fluorescence detection of head and neck tumors with 5-ALA. *Photodynamic News*, **3**(3), 8–9.

Majumder, S.K., P.K. Gupta and A. Uppal (1999). Auto fluorescence spectroscopy of tissues from human oral cavity for discriminating malignant from normal types. *Lasers Life Sci.*, **8**, 211–227.

Manoharan, R., Y. Wang and M.S. Feld (1996). Histochemical analysis of biological tissues using Raman spectroscopy. *Spectrochim. Acta A*, **52**, 215–249.

Molecular Probes Catalog (1996). In *Handbook of Fluorescent Probes and Research Chemicals*, 6th edition. R.P. Haugland (Ed.). 4849, Pitchford Ave., Eugene, OR 97402–9115, USA.

Onizawa, K., H. Saginoya, Y. Furuya and H. Yoshida (1996). Fluorescence photography as a diagnostic method for oral cancer. *Cancer Lett.*, 108, 61–66.

Panjehpour, M., B.F. Overholt, T. Vo-Dinh, R.C. Haggilt, D.H. Edwards and F.P. Buckley (1996). Endoscopic fluorescence detection of high grade dysplasia in Barrett's oesophagus. *Gasteroenterology*, **111**, 93–101.

Papazoglou, T.G. (1995). Malignancies and atherosclerotic plaque diagnosis: is laser-induced fluorescence spectroscopy the ultimate solution? *J. Photochem. Photobiol. B.*, **28**, 3–11.

Peng, Q., K. Berg, J. Moan, M. Kongshaug and J.M. Nesland (1997). 5-aminolevulinic acid–base photodynamic therapy: principles and experimental research. *Photochem. Photobiol.*, **65**, 235–251.

PharMingen (1999). Research products Catalog. San Diego, CA, USA.

Ramanujan, N., M.F. Mitchell, A. Mahadevan-Jansen, S. Thomsen, A. Malpica, T. Wright, N. Atkinson and R. Richards-Kortum (1996). Cervical precancer detection using a multivariate statistical algorithm based on laser-induced fluorescence spectra at multiple excitation wavelength. *Lasers Surg. Med.*, **19**, 63–74.

Romer, T., M. Fitzmaurice, M. Cothren, R. Richards-Kortum, R. Petras, M. Sivak and J. Kramer (1995). Laser-induced fluorescence microscopy of normal colon and dysplasia in colonic adenomas: implications for spectroscopic diagnosis. *Am. J. Gastroenterol.*, **90**, 81–87.

Scheper, T., K.D. Anders, R. Freitag, H.G. Hundeck, W. Muller, C. Schelp, A.F. Buckmann and K.F. Reardon (1989). Biosensors systems for process control in biotechnology. In *Biosensor Application in Medicine, Environment Protection and Process Control*, Vol. 13 of GBF Monographs. R.D. Schmid and F. Scheller (Eds.), VCH Publishers, New York, p. 253.

Schomacker, K.T., J.K. Frisol, C.C. Compton, T.J. Flotte, J.M. Richter, N.S. Nishioka and T.F. Deutsch (1992). Ultraviolet laser induced fluorescence of colonic tissues. Basic biology and diagnostic potential. *Laser Surg. Med.*, **12**, 63–78.

Schuimaker, J.J., H.L.L.M. van Leengoed, F.W. van der Meulen, W. Star and M. van Zandwijk (1996). Photodynamic therapy: a promising modality for the treatment of cancer. *J. Photochem. Photobiol. B*, **34**, 3–12.

Sigma Chemicals (2002). PO Box 14508, Saint Louis, MO 63178, USA.

Stummer, W., and R. Baumgartner (2000). Fluorescence-guided resection of malignant gliomas utilising 5-ALA-induced porphyrins. *Photodynamic News*, **3**(3), 6–7.

Tang, G.A., A. Prahan, W. Sha, J. Chen, C.H. Liu, S.J. Wahl and R.R. Alfano (1989). Pulsed and CW laser fluorescence spectra from cancerous, normal and chemically treated normal human breast and lung tissues. *Appl. Optics*, **28**, 2337–2342.

Van den Bergh, H., N. Lange and P. Jichlinski (1999). ALA-hexyl ester: A second generation precursors for protoporphyrin IX in photodynamic therapy and photodetection of early bladder cancer. *Photodynamic News*, **2**(1), 4–8.

Vogelstein, B., D. Lane and A.J. Levine (2000). Surfing the p 53 network. *Nature*, **408**, 307–309.

Wagnières, G.A., W.M. Star and B.C. Wilson (1998). *In vivo* fluorescence spectroscopy and imaging for oncological applications. *Photochem. Photobiol.* **68**, 603–632.

Wang, R., U. Narang, P.N. Prasad and F.V. Bright (1993). Affinity of antifluorescein antibodies encapsulated with a transparent sol–gel glass. *Anal. Chem.*, **65**, 2671–2675.

Chapter 3
ELECTROCHEMICAL BIOSENSORS

Vibha Saxena and B.D. Malhotra

OUTLINE

1. INTRODUCTION

Most strategies in analytical chemistry need complex instrumentation, special laboratory facilities and highly skilled personnel. For decades,

Advances in Biosensors
Volume 5, pages 63–100.

analytical, biological and clinical chemists have been searching for small, robust, portable, cheap, reliable and easy to use devices. The answer came in 1962 when Clark and Lyons fabricated a device called a biosensor, which utilized an oxygen electrode coated with a layer of gel containing the enzyme glucose oxidase. Following Clark's lead, others applied this technology for the fabrication of devices relating to utilization of lactate, creatinine, cholesterol, urea, uric acid and other analytes of clinical importance.

In general, biosensors comprise an analyte-selective interface in close proximity to or integrated with a transducer whose function is to relay the interaction between the surface and the analyte either directly or through a chemical mediator. The design and technology of biosensors are developing areas in the quest for innovative approaches to analysis. They have the potential to revolutionize analytical methodology by providing a powerful and often considerably less expensive alternative to older, well-established laboratory techniques. In particular, monitoring of therapeutic drug levels, office/home testing and implantable devices will benefit from the development of biosensors.

The development of catalytic biosensors can be divided into three generations characterized by the interrelationship of the functional steps – molecular analyte recognition, physicochemical signal generation and signal processing. The first generation uses single or sequentially coupled enzymes for analyte conversion to generate a transducer-accessible substance. The requirement to improve the analytical performance, i.e. specificity, sensitivity and handling co-reactants and auxiliary enzymes, led to the development of second-generation biosensors. The direct absorptive or covalent fixation of biocomponents at the transducer surface leads to the elimination of inactive membrane layers. The first biosensor of the second generation was made commercially available by Genetics International (MediSense), Abingdon, UK. This blood glucose sensor is based on the ferrocene-mediated glucose oxidation jointly developed by Cranfield and Oxford groups (Cass *et al.* 1984), which was subsequently followed by others for the measurements of other analytes. The demand for miniaturized and multifunctional biosensors in the field of neurological and physiological studies paved the way for third-generation biosensors. In this kind of biosensors the electronic signal processing unit is integrated with the sensor body. This has resulted in many successful devices such as CHEMFET (Chemical Field Effect Transistor), ENFET (Enzyme Field Effect

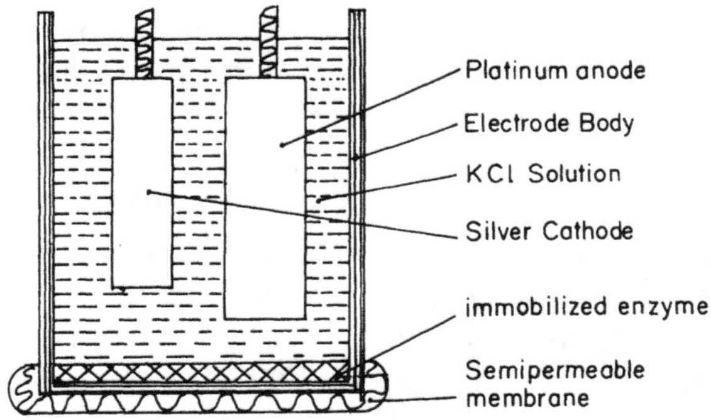

Pt anode $\rightarrow H_2O_2 \rightarrow O_2 + 2H^+ + 2e^-$

Ag Cathode $\rightarrow 2AgCl + 2e^- \rightarrow 2Ag + 2Cl^-$

Figure 1. Schematic diagram of Clark-type oxygen electrode.

Transistor), BioFET (Biological Field Effect Transistor), etc. (Koopal *et al.* 1994).

The nature of the transducer and the transduced parameter depends on the sensor type, e.g., a system designed for immunoassay wherein antibody–antigen binding must be detected is unlikely to be appropriate for an enzyme-linked redox reaction. However, since several parameters can often alter during an analytical reaction pathway, the choice of device is not necessarily restricted to a single transducer. Among the transducers, electrochemical methods are gaining increasing importance owing to their high reliability, robustness and sensitivity, their simplicity in fabrication and their low cost (Scheller and Schubert 1992). Besides this, electrochemical transducers show a large number of advantages in comparison with other systems, e.g., small amounts of analyte required, higher precision, wider linear range of measurements, absence of stray light interference, higher level of hygiene, and potential for small hand-held devices.

The evolution of the electrochemical biosensor started in 1962 (Clark and Lyons 1962). Figure 1 shows a schematic diagram of the Clark-type oxygen electrode. This arrangement allows the reduction of oxygen at the cathode to proceed. The measured cathodic current resulting from this electrochemical reduction is directly proportional to the oxygen level in the solution. Thus the depletion of oxygen at the biosensor tip due to the oxidase reaction can easily be correlated with the substrate concentration.

This process is rather complex and involves many steps of electron transfer and chemical events. Consequently, there are many parameters which influence the rate of these reactions. The overall sensor current thus depends on many factors. The device has provided a conceptual basis for the subsequent work in the development of immobilized biomembranes. The first commercial electrochemical biosensor came from Yellow Springs Instruments Company (Ohio) in 1975: a glucose amperometric biosensor. In 1987, MediSense marketed the Exactech glucose sensor (Cardosi and Turner 1991). Subsequent models based on mediated amperometry gained the major market share. I-Stat introduced a hand-held system that was capable of six clinical tests at a time with electrochemical sensors. The first Indian amperometric glucose biosensor available in the market was developed by National Physical Laboratory, India.

Table 1. Some commercial available electrochemical biosensors

Manufacturer	Analyte	Model	Range of measurement
Yellow Springs Instruments, Yellow Springs, OH, USA	Glucose lactate	23 A 23L	1–45 mM 0–15 mM
Fuji Electric, Tokyo, Japan	Glucose urate	Gluco 20 UA-300A	0–27 mM 50–60 mM
EKF Industrie-Elektronik GmbH	Glucose	Biosens 5040	–
Gamma Instrumentation India Ltd.	Glucose	Glucometer	0–600 mg/dl
Pulsatum Health Care Ltd., India	Glucose	Glucometer	0–600 mg/dl
Abott Laboratories, Abott Park, USA (formerly Medisense)	Glucose	Exactech glucose sensor	–
ABD Instruments	Glucose	ABD 3000 biosensor	–

Table 1 summarizes some commercially available electrochemical biosensors. Because of the advantages of the electrochemical biosensor, many researchers have focused towards its development for application in sample analysis and synthesis, and it has already proven itself in use. Efforts continue to increase the quality of the electrochemical response. The present chapter is focused on this aspect.

2. CLASSIFICATION OF ELECTROCHEMICAL BIOSENSORS

According to the method of detection, enzyme-based electrochemical biosensors can be classified as:

(i) Potentiometric biosensors – these use ion-selective electrodes to determine changes in concentration of chosen ions, e.g., hydrogen ions.

(ii) Amperometric biosensors – these determine the electric current associated with the electrons released during redox processes.

(iii) Conductimetric/Impedimetric biosensors – these determine conductance/impedance changes associated with changes in the overall environment.

The basic principle of the operation of these biosensors is shown in Figure 2.

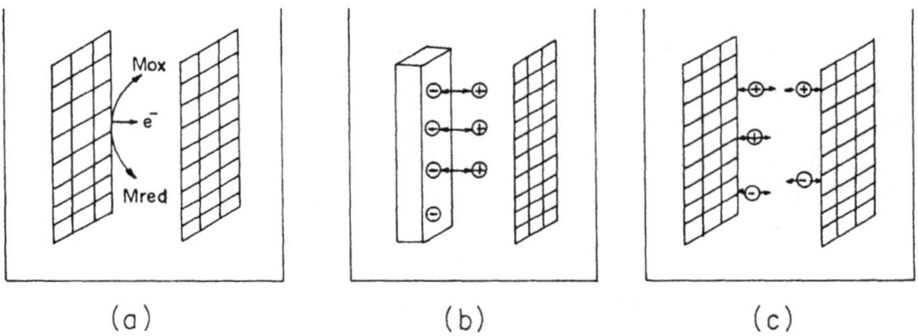

Figure 2. Principle of (a) amperometric, (b) potentiometric and (c) conductimetric biosensors; M_{red}, reduced mediator; M_{ox}, oxidised mediator; $\ominus$, anion; $\oplus$, cation.

2.1. Potentiometric Biosensors

A potentiometric biosensor consists of an ion-selective electrode (ISE) combined with immobilized enzymes. In an ISE, the electrode potential generated at a membrane electrode surface in equilibrium is proportional to the logarithm of the analyte concentration. Guilbault and Montalvo (1969) were the first to devise a potentiometric enzyme electrode for the detection of urea where urease was immobilized at an ammonia-selective liquid membrane electrode. The simplest potentiometric technique is based on the concentration dependence of the potential E, at reversible redox electrodes according to the Nernst equation:

$$E = E_0 + \frac{RT}{nF}\ln[a], \tag{1}$$

where E_0 is the standard redox potential, R is the gas constant, T is the absolute temperature, F is the Faraday constant, N is the number of exchanged electrons of substance S, and a is the activity of substance S.

The classical glass pH electrode was essentially the first ISE routinely used in analytical practice. There are innumerable reports available in the literature based on this method for measurements of desired analytes (Caras and Janata 1985, Trojanowicz and Krawczyk 1995, Pandey and Mishra 1988, Yamato *et al.* 1995). Despite their outstanding selectivity for H^+ ions, these glass electrodes are used only seldom in enzyme electrodes because of their high resistance, fragility, erosion in hydrofluoric acid and influence on sensitivity by the buffer capacity of the measuring solution. A number of other pH-sensitive materials have been investigated, including antimony (Przybyt and Sugier 1990a), tungsten oxide (Przybyt and Sugier 1990b), iridium oxide (Tarlov *et al.* 1990), stainless steel (Greenway *et al.* 1994) and conducting polymers (Pandey 1988). Miniaturization of potentiometric enzyme electrodes has been achieved using coated wire electrode. Taguchi *et al.* (1990) have reported a tridodecylamine H^+-selective neutral carrier coated over a platinum wire for the fabrication of a penicillin microelectrode.

The ISE draws on a very small current and thus necessitates a high-impedance measurement of the emf generated. The signal is, therefore, highly vulnerable to background electrical noise. The semiconducting structures now most frequently used as the transducers for biosensor fabrication were introduced by Bergveld (1970). The enzyme molecule is immobilized on the ion-sensitive surface layer of the sensing device and the resulting surface charge modulates the space charge region at the semiconductor–insulator interface. Ion-dependent changes in charge density at the gate region modulate current flow between the source and drain. The device, therefore, has a low-impedance output, and is also robust, solid-state, miniaturized and low cost (Figure 3). Other advantages are the minute amount of biological material it requirs, its multifunctionality and its rapid response. Since the introduction of such enzyme-linked FETs (Field Effect Transistors) for penicillin in 1980, FETs have been used for the fabrication of integrated biosensors (Caras

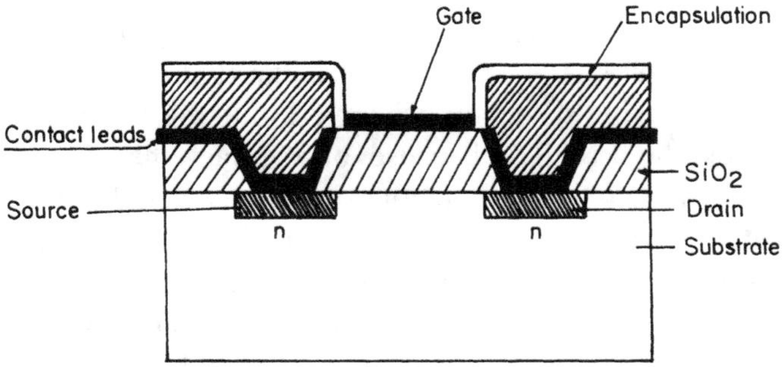

Figure 3. Schematic diagram of an ion-selective field effect transistor.

and Janata 1980). The devices based on this principle are ISFETs (ion-selective field effect transistor) (Van der Schoot and Bergveld 1987–1988, Bergveld *et al.* 1995), CHEMFETs (Blackborn 1987), ENFETs (Soldatkin *et al.* 1997), BioFET (Schutz *et al.* 1996, Schoning *et al.* 1998a), electrolyte–insulator–semiconductor (EIS) structures (Beyer *et al.* 1994, Menzel *et al.* 1995, Schoning *et al.* 1998b, Thust *et al.* 1998, Luth *et al.* 2000) and light-addressable potentiometric sensors (LAPS) (Hafeman *et al.* 1988, Owicki *et al.* 1994, Seki *et al.* 1998). In recent years, considerable efforts have been made in devising selectively modified FETs for biomedical and clinical analysis applications (Poghossian *et al.* 2001, Bergveld *et al.* 1995, Miyahara *et al.* 1985). Using this approach, devices were developed for the measurement of urea (Shiono *et al.* 1986), creatinine and penicillin (Caras and Janata 1985, Dransfeld *et al.* 1990). Enzyme inhibition has also been exploited for analysis by the pH FET, e.g., inhibition of an acetylcholinesterase layer in the presence of organic phosphate pesticide allows the detection of this agent through attenuation of the acetic acid signal from the enzymic reaction (Dumschat *et al.* 1991, Tran-Minh *et al.* 1990). This kind of approach offers high sensitivity and specificity and a wide dynamic range, which is especially useful for the determination of a wide range of toxic compounds. However, some of these ISFETs suffer from (i) physical instability and light sensitivity of semiconductor structures and (ii) decrease in response in highly buffered media. To avoid the first problem, specially designed pH FETs were proposed by Shulga *et al.* (1995). In order to reduce the effect of the buffer, the application of some additional membranes on the top of the enzymatic layer was proposed (Soldatkin *et al.* 1994, Volotovsky *et al.* 1996, Dzyadevich *et al.* 1999). This additional membrane also results in an extended dynamic range of the sensor (Munoz *et al.* 1997) and improved performance by rejecting interferents and reducing background error (Rosario *et al.* 1990). The underlying detection principle limits the number of analytes being detected by ENFETs because of the fact that construction of ENFETs are limited to the reactions involving acid–base equilibrium of the medium. A major practical problem with the manufacturing of ENFETs is the need of water-proof encapsulation of silicon (Merkoci *et al.* 1999, Zolfino *et al.* 1997). Buch *et al.* (1991) demonstrated the possibility of an intact chemoreceptor-based biosensor by using glass-capillary microelectrodes to interface the signal between a crab antennule and electronics. However, this kind of approach could not be miniaturized. To overcome the problem, a direct 'silicon–neuron' junction was implemented by Fromherz *et al.* (1991, 1993). The immobilization of a single neuron to the gate of a FET allowed the first direct bioelectronic signal transfer. Schutz *et al.* (1997) combined both approaches and developed a novel type of BioFET consisting of a FET–insect antenna junction. Using this approach, several BioFETs have been developed with the possibility of applications in agriculture or fire detection (Schutz *et al.* 1996, Schoning *et al.* 1998a, Kaissling 1995).

The light-addressable potentiometric sensor (LAPS) exploits the H^+ sensitivity of silicon nitride and thus avoids protective encapsulation of the semiconductor. In this transducer, specific locations on a sensor are made sensitive to pH by illumination from a LED array. The method has been used to develop biosensors for the detection of clinically important analytes, pathogenic bacteria, anticholinesterase drugs and pesticides and biological warfare agents (Uithoven *et al.* 2000, Hafeman *et al.* 1988, Ivnitski *et al.* 1999). Another notable report is the use of a catalytic antibody, which combines with the signal-generating capability of an enzyme (Blackburn *et al.* 1990). The possibility for reversible antigen attachment could also make continuous reversible operation possible, allowing unprecedented continuous monitoring of an antigen.

2.2. Amperometric Biosensors

Amperometric biosensors have been at the focus of electroanalytical research since the first report of the enzyme electrode by Updike and Hicks in 1967 for the detection of glucose. The amperometric transducer measures current resulting from the oxidation or reduction of electroactive compounds. The current is proportional to the concentration of the analyte. The first-generation device of this kind was made by using either an oxygen electrode or a hydrogen peroxide electrode, for instance,

$$\text{substrate} + O_2 \xrightarrow{\text{enzyme oxidase}} \text{product} + H_2O_2 \tag{2}$$

or

$$H_2O_2 \xrightarrow{+650\,\text{mV vs Ag/AgCl}} O_2 + 2H^+ + 2e^-. \tag{3}$$

The electrode measures the analyte concentration by determining either the consumption of oxygen or the production of hydrogen peroxide. Based on this principle many substrates such as glucose, lactate (Guilbault and Lubrano 1973), ascorbate (Daily *et al.* 1991), oxidase (Peguin *et al.* 1989), etc., have been assayed. There were some serious problems with these enzyme electrodes: the oxygen electrode is susceptible to oxygen variation in the sample and the hydrogen-peroxide-based electrode has poor selectivity when operating in biological fluids: most such fluids encounter interferents that are electrochemically active at the high polarizing voltages needed for hydrogen peroxide detection. In blood, thiols, ascorbate, urate and drugs can cause interference during clinical measurements. The alternative is to use diffusion-limiting external covering membranes such as cellulose acetate, polyurethane, Nafion, and Eastman AQ-29 D. Several papers have reported the use of membranes for oxygen electrodes (Keedy and Vadgama 1991, Tang *et al.* 1990, Bindra *et al.* 1991) and permselective membranes for hydrogen peroxide (Vadgama *et al.* 1989, Gorton *et al.* 1990).

Another alternative to circumvent the problem of oxygen and hydrogen peroxide electrodes is to use mediators which transfer electrons directly to the electrode, bypassing the reduction of the oxygen co-substrate. Figure 4 shows a schematic picture of electron transfer in a mediated system. These mediators should possess the following properties:

(i) They must react rapidly with the reduced form of enzyme.
(ii) They must be sufficiently soluble in both reduced and oxidized forms.
(iii) Overpotential for the regeneration of the oxidized species at the electrode should be low and independent of pH.
(iv) The reduced form of the mediator should not react with oxygen.

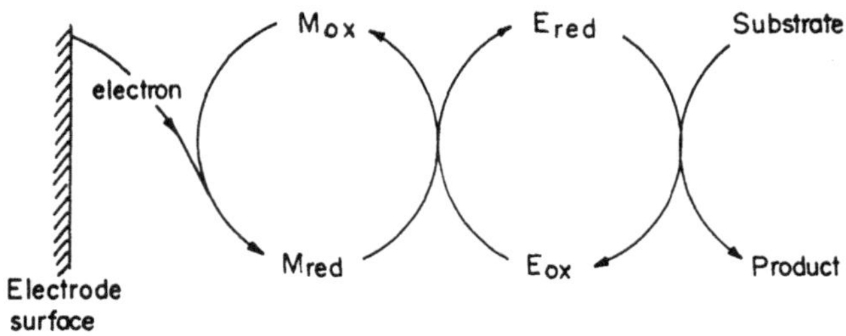

Figure 4. Schematic picture of electron transfer in a mediated biosensor. M_{red}, reduced mediator; M_{ox}, oxidised mediator; E_{red}, reduced enzyme; E_{ox}, oxidised enzyme.

Many electron acceptor molecules and complexes have been used as mediators. Among these, ferrocene and its derivatives, NADH, benzoquinone, methylene blue, have attracted the most attention (Katakis and Dominguez 1997, Lobo *et al.* 1997). The advantage of these approaches is that there are a large numbers of enzymes selective for particular compounds that can be employed in an amperometric biosensor, provided there is an appropriate cofactor that is electroactive in an electrode potential region that yields a low background current. NADH is often employed as a cofactor in a large number of dehydrogenases. Its direct oxidation at the bare electrode surface often requires large overpotentials (Moiroux and Elving 1978, Jaegfeldt 1980), which can be brought down through redox mediators such as thionine derivatives (Hahizadeh *et al.* 1991), phenazine (Torstesson and Gorton 1981), phenoxazine (Persson and Gorton 1990), derivatives absorbed onto graphite (Huan *et al.* 1996), or electrodeposited films (Persson *et al.* 1990).

The properties of the enzyme-based amperometric biosensor are predominantly governed by the ability of the immobilized enzyme to exchange electrons with the electrode at an appreciable electron transfer rate. Methods such as cross-linking, physical trapping and covalent linking are all traditional (Guilbault 1984, Wilson and Thévenot 1989). Recent research has promoted the interest in the utilization of some novel methods, such as sol–gel (Niu and Lee 2000), electropolymerized film (Foulds and Lowe 1988, Garjonyte and Malinauskas 2000) and redox-relay modified hydrogels (Degani and Heller 1989, Danilowicz *et al.* 1999). A promising immobilization method is the entrapment of the proteins/enzyme in conducting polymer films. Advantages of using this approach are (i) an easy one-step preparation method, (ii) control of film thickness and polymerization parameters (Fuji *et al.* 1990, Lin and Wallace 1994, Kranz *et al.* 1998) and (iii) localization of the electrochemical reaction exclusively on the electrode surface (Yon Hin *et al.* 1990) and the possibility of fabricating multilayer structures (Bartlett and Birkin 1993). Thus intensive research has been focused on using biosensors based on conducting polymer (Bartlett and Whitaker 1987, Pandey and Mishra 1988, Deshpande and Amalnekar 1993, Schuhmann *et al.* 1993, Trojanowicz and Krawczyk 1995, Cosnier 1999). The most commonly used conducting polymers are polyaniline (Eftekhari 2001, Ramanathan *et al.* 1995, Chaubey *et al.* 2000) and polypyrrole (Gaspar *et al.* 2001, Ramanathan *et al.* 1996), which have been used to detect biologically important species. The advantages of redox hydrogel-based biosensors were combined with those of the electrochemical formation of conducting polymer films by modifying polymer with covalently bound ferrocene derivatives (Foulds and Lowe 1988) or osmium complexes (Schuhmann *et al.* 1993). However, only slow electron-transfer rates resulting in low current densities could be obtained due to hydrophobicity and rigidity of the polymer backbone. Recently, it was demonstrated that a significant increase in current could be obtained by modulating the hydrophilicity of the conducting polymer backbone by using a PQQ-dependent glucose dehydrogenase, thereby improving the electron-transfer kinetics within the redox polymer film (Habermuller *et al.* 2000, Reiter *et al.* 2001). For more details on biosensors based on conducting polymer, we refer the reader to Chaubey and Malhotra (2002).

New promising trends in the development of enzyme electrodes base on chemically modified electrodes, e.g., organic-phase enzyme electrodes (OPEEs). OPEEs contain redox enzymes as the biological sensing element that are capable of entering into a redox catalytic reaction with the analyte in an organic solvent environment, and are thereby useful in operating in organic media as well. Consequently, OPEEs are able to analyse almost any compound. Some of the reports available in literature include detection of phenols (J. Wang *et al.* 1993), peroxides (Adeyoju *et al.* 1994, Schubert *et al.* 1992), pesticides (Adeyoju *et al.* 1995) and metabolites (Hall and Turner 1991).

A recent and growing trend in the design of amperometric biosensors involves the incorporation of a biocatalyst within the bulk of a carbon composite matrix (Gorton 1995, Wang *et al.* 1993). The resulting bulk-modified bioelectrode offers several advantages: (i) the close proximity of the biocatalyst and sensing sites, (ii) the possibility to incorporate other components, e.g., cofactors or redox mediators, (iii) an easy renewal of the surface, (iv) easy and economic fabrication, and (v) a high sensitivity of the incorporated biocatalysts. Examples of these bioelectrodes are carbon paste electrodes (Gorton 1995), solid composite biosensors using various solid binders, like epoxy matrix (Alegret *et al.* 1996) or ceramic (Sampath and Lev 1996), and solid binding matrices of defined molecular structure (Svorc *et al.* 1995).

2.3. Conductimetric/Impedemetric Biosensors

The general principle underlying conductivity/impedance-based biosensors is the observation of change in conductivity/impedance as a result of change in concentration of some ionized species in an enzyme reaction. Though conductance is determined by the migration of all ions present in the solution, substrate selectivity is achieved via the specific nature of the enzyme reaction. Conductimetric biosensors are widely recognized as an attractive basis for biosensors due to their direct electrical response, non-requirement of reference electrode, possibility of fabrication by thin-film technology and, therefore, good compatibility with electronic circuits and computer interfaces. Watson *et al.* (1987) integrated an enzyme with a conductimetric detector by cross-linking urease over gold interdigited electrodes that were microfabricated on a silicon wafer. A subsequent paper (Yon Hin *et al.* 1990) reported the fabrication of a monolithic device using an electrochemically synthesized polypyrrole film. Since conductimetric response is partly due to the change in H^+ concentration, both sensitivity and response time can be expected to be a function of sample pH and buffer capacity. This was observed by Mikkelsen and Rechnitz (1989) who devised a conductimetric D-amino acid sensor using an oxidase. The conductimetric method was also used in measuring time-dependent changes of conductivity in microbial biomass. Mostly, conducting polymers and phthalocyanines were used as materials for the transducing element. Palmqvist *et al.* (1994) reported a device based on conducting polypyrrole, which functioned in an aqueous phase and required very simple measuring equipment. Fare *et al.* (1998) studied the properties of poly(3-hexylthiophene)-coated Pt electrode and developed a sensor for immunoassay read-out based on admittance measurements. Kim *et al.* (2000) investigated an immunochromatographic assay system based on conductimetric detection for point-of-care examination by utilizing, as a signal generator, colloidal gold with polyaniline bound on the metal surface. This novel signal generator amplified the conductimetric signal 4.7 times compared with plain gold, and the signal was also maximum, 2.3-fold higher than that from the photometric system under the same analytical conditions. One major drawback of conducting polymers is that they are unstable in aqueous solution. Palmqvist *et al.* (1995) reported a biosensor that employs a dry conducting polypyrrole film exposed to a gas phase, thereby offering advantage in long-term stability. Sergeyeva *et al.* (1998) reported a new approach to conductometric biosensors utilizing iodine-sensitive phthalocyanine thin films. The sensitivity of the method proposed was high enough to detect unlabelled IgG in blood serum in competitive mode. There are only a few reports on the successful function of biosensors in truly conductimetric mode (Bataillard *et al.* 1988, Shulga *et al.* 1994, DeSilva *et al.* 1995), while considerably more advances and practical achievements have been obtained in the field of amperometric and potentiometric biosensors (Turner *et al.* 1987, Hall 1990). The main reason is that the relatively low impedance of aqueous media makes it difficult to operate conductimetric biosensors.

Capacitance measurement techniques can also be used to develop a biosensor. Capacitance measurements on biosensors may succeed only if the successive biomolecular layers grafted onto the heterostructures are sufficiently electrically insulating and retain their recognizing ability. However, Berney *et al.* (1997) demonstrated that it is possible to develop a sensor if the biolayer is not sufficiently insulating or does not have a suitable dielectric character. They used an overlay of non-conducting polymer, in the form of polyethylene glycol (PEG) and showed the possibility of developing differential capacitive biosensors. There is also considerable literature on voltammetric (Fernandez-Sanchez *et al.* 2000) and impedimetric biosensors (Pak *et al.* 1998).

3. APPLICATIONS OF ELECTROCHEMICAL BIOSENSORS

3.1. Clinical and Diagnostics

Electrochemical biosensors have found wide interest in the clinical diagnostics market. Many attempts have been directed to the commercialization of these biosensors. To date, many biosensors are available commercially for the detection of glucose, lactate, uric acid, cholesterol, creatinine, etc.

Glucose has been the most studied analyte in diabetic patients. The glucose level can be monitored either *in vivo* or *in vitro*. The first approach for *in vivo* study was pioneered by Shichiri *et al.* (1986). They implanted needle sensors subcutaneously for several days. The current signals generated were converted to very-high-frequency (VHF) audio signals. One of the best-known portable biosensors for home-monitoring of glucose for diabetics has been marketed by Medisense Inc. (Cambridge, USA). This biosensor is based on screen-printed technology and employs glucose oxidase in a pen-shaped device (Cass *et al.* 1984). The main disadvantages of this pen are, first, that its glucose strips cannot be reused and, second, that it cannot be applied in continuous monitoring. Keeping this in view, Mascini *et al.* (1987) reported an "artificial pancreas" for continuous measurement of glucose. Another commercially available biosensor is the I-STAT portable clinical analyser that can measure a range of parameters: sodium, chloride, potassium, glucose, blood, urea, nitrogen and haematocrit. The sensors are fabricated using thin-film microfabrication technology on a disposable cartridge (Erickson and Wilding 1993). Advances to achieve the goal of techniques suitable for whole-blood immunoassay were made by Molecular Devices Inc., with their light-addressable potentiometric sensor. The system has been used to assay total DNA (Genesis report 1992).

Another metabolite important in clinical diagnostics is L-lactate whose measurement is helpful for monitoring respiratory insufficiency, shocks, heart failure, metabolic disorder, tissue injuries, thrombosis and the physical condition of athletes. Many electrochemical biosensors have been reported to date (Dempsey *et al.* 1997, Chaubey *et al.* 2000). Palmisano et al. (1994) determined L-lactate using a platinum electrode with an electropolymerised poly(o-phenylenediamine)/lactate oxidase film. Palleschi and Turner (1990) made tetrathiafulvalene-mediated lactose electrodes using lactate oxidase adsorbed on carbon foil. Tatsuma and Watanabe (1991) used oxidase/peroxidase bilayer-modified electrodes to detect L-lactate. An osmium bilayer hydrogel was used for wiring lactate oxidase to vitreous carbon electrodes by Ohara *et al.* (1994). Interest in a miniaturized lactate sensor has been rising. Two different technologies have been used. Firstly, thin-film electrodes were developed which can be used either as implantable catheter-type devices or for *in-vivo* monitoring in combination with a microdialysis system (Pfeiffer *et al.* 1997, Tatsuma and Watanabe 1991). Secondly, disposable sensors were developed for the purpose of on-line analysis (Patel *et al.* 2000, Hart *et al.* 1999). Recently, an ultrafiltration catheter has been reported for monitoring absolute lactate and glucose concentrations (Renger *et al.* 2001).

Creatinine is another important clinical analyte, used for the determination of renal and muscular dysfunction. Attempts to fabricate a potentiometric device began in 1976 with Meyerhoft and Rechnitz. Later on, Guilbault and Coulet (1983) developed a similar system with improved operational and storage capability. However, the system fell far short of commercial requirements due to its limited working range. Mascini *et al.* (1985) extended the operation and storage stability of these devices and reached the analytical range required for reporting normal as well as pathological specimens. Up to date, all potentiometric devices have been based on the catalysis of creatinine by creatinine hydrolase (CIH) at the surface of an NH_4^+-sensing ion-selective electrode (Shin *et al.* 1998, Osaka *et al.* 1998). The device fabrication is rather simple, but significant problems are caused by interference from endogenous NH_4^+ in blood and urine samples. Recently, pre-analytical removal of endogenous ammonia has been reported, which uses thermal decomposition of alkalinized urine samples with no detrimental effect on creatinine detection (Shin and Huang 1999). However, the introduction of additional pre-analytical steps is undesirable from a commercial point of view. Recently, an impedimetric device has been reported to assay urea and creatinine in serum using screen-printed electrodes modified by poly(methylvinyl ether)/maleic anhydride (Ho *et al.* 1999). Among the electrochemical methods used for the determination of this analyte, amperometry has proved most useful. Almost all amperometric biosensors have been based on a three-enzyme method described by Tsuchida and Yoda (1983), which involves the three-stage conversion of creatinine to creatine, creatine to sarcosine and sarcosine to glycine. In the final stage, electrochemical consumption of oxygen and liberation of hydrogen peroxide (H_2O_2) occurs. Detection of H_2O_2 liberation is the preferred technique in amperometric systems (Schneider *et al.* 1996, Madaras and Buck 1996, Khan and Werner 1997). The first commercially successful system was from Nova Biomedical (USA). In addition, Abott Inc. has incorporated creatinine detection into their I-STAT point-of-care system.

Determination of cholesterol is clinically very important because abnormal concentrations of cholesterol are related with hypertension, hyperthyroidism, anemia and coronary artery diseases. Determination based on the inherent specificity of an enzymatic reaction provides the most accurate means for obtaining true blood cholesterol concentration. The direct electrochemical measurement of thiocholine iodide has been used for this purpose by Gruss and Scheller (1987). Amperometric cholesterol sensors were prepared by immobilizing cholesterol oxidase (ChOx) on the electrode surface, and the enzymatic reaction between ChOx and cholesterol was transduced into a current signal using electron-transfer mediators such as metal complexes and phenothiazine dyes (Gilmartin and Hart 1994, Motonaka and Faulkner 1993, Yon Hin and Lowe 1992, Nakaminami *et al.* 1997). Electrochemical determination of biological compounds suffers severe interference in the presence of compounds such as L-ascorbic acid and acetaminophen present in physiological samples. Several strategies have been used for the modification of electrode surfaces. Conducting polymers, e.g. polypyrrole (Yon Hin and Lowe 1992) and polyaniline (Haiyan and Shaolin 1999), have been used for this purpose. Further improvements in the stability and selectivity were achieved by using either a bilayer (Vidal *et al.* 1999) or a sol–gel (Yao and Takashima 1998). The self-assembly technique is of particular interest because of easier fabrication, ordered arrangement and high reproducibility. Recently, Gobi and Mizutani (2001) demonstrated an amperometric biosensor for cholesterol determination by a layer-by-layer nano thin film formation using ChOx and poly(styrenesulfonate) on a monolayer of microperoxidase covalently immobilised on Au–alkanethiolate electrodes. The sensor was found to be responsive even in the presence of potential electrical interferents, L-ascorbic acid, pyruvic acid and uric acid.

The measurement of metabolites in media other than blood is becoming increasingly significant because of major demands for non-invasive analysis. Such measurements are particularly important for patients who have to control daily parameters such as glycemia and urea and for people having problems in collecting blood. Several electrochemical biosensors have been developed and applied for non-invasive determination of glucose (Pallechi *et al.* 1991), lactate (Faridnia *et al.* 1993), lactic acid (Pilloton *et al.* 1988), alcohol (Jones 1979, Guilbault and Palleschi 1995), etc., in saliva or sweat. Table 2 lists literature available on electrochemical biosensors for the detection of clinically important analytes.

Table 2. Some electrochemical biosensors for detection and estimation of analytes

Analyte	Biosensor type	Lower limit of detection	Assay time	Reference
Ochratoxin	Amperometric	10^{-4} ppm	–	Albery *et al.* (1990)
Cyanide	Amperometric	0.01 ppm	Steady state within 1 min	Aizawa (1987)
Lactate	Amperometric	0.1 mM	1.2 min	Nguyen and Luong (1993)
Alcohol	Amperometric	0.2 mM	30 s	Kulys and Schmid (1991)
Urea	ENFET	0.1 mM	95% of steady state within 1 min	Munoz *et al.* (1997)
Urea	Conductimetric	–	20 min	Palmqvist *et al.* (1995)
Uric acid	Amperometric	0.1 mM	20 s	Nakaminami *et al.* (1999)
Glucose	Amperometric	–	<20 s	Pandey and Upadhyay (2001)
Alcohol	Amperometric	0.1mM	–	Mullor et al (1996)
Phenol	Amperometric	5×10^{-7} M	–	Kaisheva *et al.* (1997)
Urea	Potentiometric	2.0×10^{-5} M	1–2 min	Eggenstein *et al.* (1999)
Glucose	Amperometric	0.1 mM	3–30 s	Puig-Lleixa *et al.* 2001
Alcohol	Amperometric	0.02 mM	~20 s	Patel *et al.* (2001)
Cholesterol	Amperometric	2 mM	30 s	Kumar *et al.* (2001)

3.2. Food and Drinks Industry

Quality control is of the utmost importance in food technology. The food and drinks industries need rapid and affordable methods to determine compounds affecting nutrition value and quality of food. Although the food industry is very large, the profit margin is low and therefore there has been some reluctance to invest in expensive analyte equipment such as gas chromatography, spectrophotometry, high-performance liquid chromatography, etc. Another disadvantage of these conventional methods is that they cannot be used for on-line monitoring. The use of biosensors to monitor freshness of food and concentration of food ingredients is a relatively recent development (Scott 1998). The major advantages of biosensors in this field are rapid response, simple operation, high specificity, low cost, non-requirement of sample pre-treatment, small size and on-line monitoring. Membrane- and electrode-based electrochemical biosensors have been used extensively in the determination of food ingredients, such as carbohydrates, sugar and starch, and monitoring fish freshness, beverages and drink. Emneus and Gorton (1990) presented a flow-injection system with a multi-enzyme biosensor for the detection of starch. The device was based on consecutive reaction steps and amperometric detection at a chemically modified electrode. The analysis of sucrose is of great importance in the food and drinks industries. A biosensor using invertase, mutarotase and glucose oxidase coupled to an oxygen sensor has been described. However, this system was sensitive to oxygen tension of the reactor and resulted in poor selectivity. Pandey *et al.* (1993) reported a device based on incorporation of enzymes into modified graphite paste for very sensitive probing of enzyme-catalysed reaction. Recently, they reported a flow injection analysis biosensor with improved sensitivity. The system was capable of detecting samples containing high concentrations of sucrose up to 2 mol/L without the requirement of sample dilution (Lima Filho *et al.* 1996). Measurement of L-lactate in food chemistry is useful for evaluating freshness and stability of milk, dairy products, fruits, vegetables, sausages and wines, and several biosensors have been developed for the purpose (Collier *et al.* 1996, 1998).

The accurate measurement of alcohol is very important in pulp, food and drinks industries. The enzymes used for ethanol detection are alcohol dehydrogenase (ADH) and alcohol oxidase (AOX). ADH has the advantage of being more stable and more specific but the disadvantage of being dependent on the dissolved coenzyme NAD^+ which has to be added to the assay. The problem was overcome by using mediators like methylene green (Chin and Dong 1994), $[Re(phen\text{-}dione)(CO)_3Cl]$ and $[Fe(phen\text{-}dione)_3](PF_6)_2$ (Tobalina *et al.* 1999), medola blue (Mullor *et al.* 1996) and 3,4-dihydroxybenzaldehyde (Tobalina et al. 1998). Wang et al. (1995) used screen-printing carbon ink with dispersed ruthenium and thus decreased the overpotential for ADH ethanol sensors. Screen-printed electrodes have been found advantageous for mass production (Park *et al.* 1995, 1999). Sprules *et al.* (1996) measured ethanol in gin using medola-blue modified single-use screen-printed biosensors. Using AOX, an ethanol biosensor was first developed by Nanjo and Guilbault in 1975. Stability has been achieved by using hydrophobic semisolid matrices (Liu and Wang 1999), enzyme reactors (Varadi and Adanyi 1994), polyion complexes on glassy carbon electrodes (Mitzutani *et al.* 1998), and by the addition of lactitol and a positively charged dextran derivative (Gibson *et al.* 1992). Recently, a low-cost screen-printed enzyme sensor was introduced comprising a polyester substrate, a working electrode made of thick-film polymer paste containing platinum, a carbon counterelectrode and an Ag/AgCl pseudo-reference electrode (Erlenkotter *et al.* 2000, Patel *et al.* 2001) for the detection of alcohol in batch mode.

The polyphenols are known for their anti-oxidative behaviour which results in better storage of olive oils and in an improvement of the nutritional features. The suitability of biosensors for monitoring phenol and polyphenols has been demonstrated for both food and environmental analysis (Romani *et al.* 2000, Macholan and Schanel 1977). Onnerfjord *et al.* (1995) reported an amperometric biosensor based on the enzyme tyrosinase, which catalyses the oxidation of phenols to the quinonic form. This is then reduced at the graphite surface polarized at −200 mV versus a reference electrode (Ag/AgCl). In addition, several biosensors have been developed for the detection of pesticides in food (Brooks and Roberts 1999, Yulaev *et al.* 1995), simazine in meat extracts, cucumber and milk (Yulaev *et al.* 2001), and histamine and other biogenic amines (Tombelli and Mascini 1998).

3.3. Environmental Monitoring

Sensor systems of this ilk present devices that are able to monitor specific environmental contaminants with the intention of providing on-site analysis and removing the time delay and potential for sample modification associated with centralized laboratory testing. The detection of a range of organic compounds such as pesticides, organohalide solvents, hydrocarbons, bacteria, toxic gases and hydrocarbons and metals in water, soil and air is currently the prime area of research (Shriver-Lake and Ligler 1995).

The measurement of biochemical oxygen demand (BOD) is important in determining water quality: BOD is the amount of oxygen required for biodegrading organic compounds in aqueous solutions. Conventional methods are time consuming and therefore unsuitable for process control. The use of a Clark electrode is a highly selective method for water quality control (Tarasevich *et al.* 1993, Jing *et al.* 2000). Another approach is based on the application of the enzyme peroxidase (Bogdanovskaya *et al.* 1994), immobilized yeasts (Hikuma *et al.* 1979) and activated sludge (Liu *et al.* 2000).

Lately, electrochemical biosensors have been employed for detection and monitoring of gaseous substances in air such as phenol, formaldehyde and ethanol. Phenolic compounds are among the major pollutants in industrial waste. Formaldehyde is classified as a mutagen and possible carcinogen (Metzger *et al.* 1998, Cosnier *et al.* 1999, Alatorre and Bedioui 1999). Using a microelectrode system consisting of two sets of golden microbands, O'Sullivan *et al.* (1996) developed biosensors for the detection of SO_2, phenol, ethanol and formaldehyde. Using a similar pattern, a biosensor for phenol with high sensitivity was developed by Dennison *et al.* (1995). High sensitivity for electrochemical detection is destroyed both by the large overvoltage required for direct oxidation of phenols and by the relative humidity of air. To overcome the first problem, modified electrodes, e.g. Clark oxygen electrode (Macholan and Schanel 1977, Macholan 1990, Campanella *et al.* 1993), solid graphite electrode (Cosnier and Innocent 1993, Ortega *et al.* 1993), carbon electrodes (Bonakdar *et al.* 1989) and a graphite-epoxy-based composite electrode (Onnerfjord *et al.* 1995), were utilized for measurements of phenols. The second problem has been overcome by using a biosensor comprising of an enzyme/gas diffusion electrode with tyrosinase enzyme. The electrode was capable of detecting phenol in the ppb range (Kaisheva *et al.* 1997).

Many approaches for the detection of formaldehyde have been published, including a system operating in gas and organic phases (Hammerle *et al.* 1996, Vianello *et al.* 1996). A number of electrochemical methods such as amperometric (Winter and Cammann 1989, Hall *et al.* 1998), potentiometric (Korpan *et al.* 1993, 1997, 2000) and voltammetric (Chan and Xie 1997), have been demonstrated successfully for the determination of formaldehyde. However, there are some serious problems restricting wide commercial application of these biosensors. For formaldehyde-dehydrogenase-based biosensors such difficulties arise from the necessary addition of cofactor enabling formaldehyde conversion (Winter and Cammann 1989) and of electrochemical mediators (Hall *et al.* 1998). Recently, Korpan *et al.* (2000) reported a highly selective and stable pH FET device for formaldehyde determination using alcohol oxidase (AOX) or permeabilised yeast cells containing AOX, which can be used as control devices in production units for direct monitoring of wastes, ambient air and water after a sample pre-concentration.

The accurate determination of organophosphatase in air and water is essential for monitoring the level of a pesticide. The most prevalent biosensor strategy is to use acetylcholinesterase (AChE) or butylcholinesterase (BChE) as the sensing recognizing component (Trojanowicz and Hitchman 1996). In an AChE-based biosensor, the signal is proportional to the organophosphate concentration. In recent years, another enzyme, organophosphate hydrolase, has been proposed by researchers as an alternative recognition component (Dave *et al.* 1993, Rainina *et al.* 1996) which leads to the direct determination of analyte. Improved device performance was achieved by using a pH-sensitive FET. The small gate area available for enzyme immobilization limits the amount of biocomponent to be immobilized. This can be achieved by increasing the surface area. For this purpose a sol–gel process was used to create a porous silica structure (Singh *et al.* 1999, Flounders *et al.* 1999, Pandey *et al.* 2000).

In addition, electrochemical biosensors have effectively been used for the measurement of formate (Yuan *et al.* 2000, Adeloju and Yuan 1999), heavy metals concentration (Palchetti *et al.* 1999), hazardous and toxic materials (Shriver-Lake and Ligler 1995), cyanide (Smit and Cass 1990, Smit and Rechnitz 1993, Hu and Leng 1995, Amine *et al.* 1995), toxins (Rogers *et al.* 1991, 1992, Colston *et al.* 1993, Aizawa 1987, DeSilva *et al.* 1995, Dill *et al.* 1994) and pesticides (Fernandez-Alba *et al.* 2001).

3.4. DNA Biosensor

In recent years there has been an increase in the use of nucleic acids as tools in the recognition and monitoring of many compounds of analytical interest in the field of environmental control, the food industry and the clinical laboratory. Besides this, detection of specific DNA sequences is important because 4000 inherited diseases are known and much effort is needed to identify the underlying mutations. Nucleic acid layers combined with electrochemical transducers produce a new type of affinity biosensor for low-molecular-weight molecules. The activity in this direction has centered upon the explanation of molecular interactions between the surface-linked DNA and the target pollutants or drugs, in order to develop devices for a rapid screening of these compounds. Electrochemical biosensors offer several advantages like easier immobilization of DNA layer (i.e. adsorption driven by a positive electrode potential), faster and cheap measurements, and the use of miniaturized and portable instrumentation (Chen and Shenghui 2000, Palanti *et al.* 1996, Marraza *et al.* 1999a).

Conventional methods for the analysis of specific gene sequences are based on either direct sequencing or DNA hybridization (Marraza *et al.* 1999b). Because of its simplicity most of the traditional techniques in molecular biology are based on hybridization. Several immobilization techniques such as adsorption (Wang *et al.* 1997), covalent attachment (Millan *et al.* 1994), or immobilization involving avidin–biotin complexation (Cosnier *et al.* 1998) were adopted for a DNA probe to the surface of an electrochemical transducer. The transducer was made from carbon (Millan and Mikkelsen 1993), gold (Hashimoto *et al.* 1998, Steel and Herne 1998, Maruyama *et al.* 2001), platinum (Moser *et al.* 1997) or conducting polymer (Livache *et al.* 1994, Korri-Youssoufi *et al.* 1997). In the case of a common sandwich assay the signal-generating species is an enzyme such as horseradish peroxidase (Nikiforov and Rogers 1995). However, for most of the applications the amplification by enzyme reaction is not high enough. Polymerase chain reaction (PCR) can overcome this problem by amplifying the sample DNA up to a range which can be detected by optical or amperometric measurement methods (Mullis and Faloona 1987, Marraza *et al.* 2000). The complementary structure of DNA offers a precise biorecognition system suitable for biosensors of affinity type. Hall *et al.* (1994) used the difference in electrochemical behaviour of ss-DNA and hybridized DNA to measure the signal. Using peptide nucleic acid probes, the intercalation of a redox-active molecule in hybridized DNA was detected by cyclic voltammetry (Millan et al. 1994) or by chronopotentiometry (Wang *et al.* 1996). The attachment of the ss-DNA molecule by a 5′ or 3′ end results in a higher efficiency for the hybridization reaction (Mikkelsen 1996, Sun *et al.* 1998). Another possibility in designing a DNA biosensor is the use of magnetic particles. Lund *et al.* (1988) linked the tagged DNA to the surface of the microsphere using a suitable reagent. Another effort is the use of a microfabrication system and micromechanical technology to the preparation of DNA samples and their analysis, e.g., DNA chip. Very dense arrays of short-stranded DNA probes have been synthesized in situ on the chip using photolithography by McGall *et al.* (1987). The immobilization of avidin–biotin systems on carbon surfaces has been realized using SECM (scanning electrochemical microscopy) by Nowall *et al.* (1998).

4. CONCLUSIONS

As discussed in this chapter, electrochemical biosensors have been employed successfully in the clinical industry. Attempts are being made to exploit their advantages in the environmental and food industry as well. Recent advances in nanofabrication technology have made it possible to integrate nanofabrication technology with molecular biology. Using electron-beam lithography and advanced resist technology, it is now possible to fabricate electrochemical biosensors that can detect and manipulate single biomolecules. By adopting the latest advances in fabrication technology it has now become possible, at least in principle, to develop molecular-level electrochemical biosensors. In spite of this progress, much research is needed in the direction of development of molecular-level electrochemical devices.

ACKNOWLEDGEMENTS

The authors are thankful to Dr. Krishan Lal, Director, NPL, New Delhi, India, for his interest in this work. Vibha Saxena is thankful to CSIR for the award of a Research Associateship.

REFERENCES

Adeloju, S.B., and Y.J. Yuan (1999). Polypyrrole-based flow amperometric biosensor for formate. http://www.uws.edu.au/cerat/news/spring99/polypyrrole.html.

Adeyoju, O., E.I. Iwuoha and M.R. Smyth (1994). Determination of kinetic parameters for the inhibitory effects of organic sulfides on an amperometric peroxide biosensor in non-aqueous media. *Talanta*, **41**, 1603–1608.

Adeyoju, O., E.I. Iwuoha and M.R. Smyth (1995). Reactivities of amperometric organic phase peroxidase modified electrodes in the presence and absence of thiourea and ethylenethiourea as inhibitors. *Anal. Chim. Acta*, **305**, 57–64.

Aizawa, M. (1987). Immunosensors. *Philos. Trans. R. Soc. London B*, **316**, 121–134.

Alatorre, O.A., and F. Bedioui (1999). The electrocatalytic reduction of organohalides by myoglobin and hemoglobin in a biomembrane-like film and its application to the electrochemical detection of pollutants: new trends and discussion. *Sensors Actuators B*, **59**, 128–133.

Albery, W.J., A.E.G. Cass and Z.X. Shu (1990). Inhibited enzyme electrodes part 2. The kinetics of the cytochrome oxidase system. *Biosensors Bioelectron.*, **5**, 379–395.

Alegret, S., F. Cespedes, E. Martinez–Fabregas, D. Mortorell, A. Morales, E. Centelles and J. Munoz (1996). Carbon-polymer biocomposites for amperometric sensing. *Biosensors Bioelectron.*, **11**, 35–44.

Amine, A., M. Alafandy, J.M. Kauffmann and M. Novak Pekli (1995). Cyanide determination using an amperometric biosensor based on cytochrome oxidase inhibition. *Anal. Chem.*, **67**, 2822–2827.

Bartlett, P.N., and P.R. Birkin (1993). The application of conducting polymers in biosensors. *Synth. Met.*, **61**, 15–21.

Bartlett, P.N., and R.G. Whitaker (1987). Electrochemical immobilization of enzymes. Part 1. Theory. *J. Electroanal. Chem.*, **224**, 27–35.

Bataillard, P., F. Gardies, N. Renault-Jaffrezic, C. Martlet, B. Colin and B. Mandrand (1988). Direct detection of immunospecies by capacitance measurements. *Anal. Chem.*, **60**, 2374–2379.

Bergveld, P. (1970). Development of an ion-sensitive solid state device for neurophysiological measurements. *IEEE Trans. Biomed. Eng.*, **BME-17**, 70–71.

Bergveld, P., R.E.G. Van Hall and J.C.T. Eigkel (1995). The remarkable similarity between the acid base properties of ISFET and proteins and the consequences for the design of ISFET biosensors. *Biosensors Bioelectron.*, **10**, 405–414.

Berney, H., J. Alderman, W. Lane and J.K. Collins (1997). A differential capacitive biosensor using polyethylene glycol to overlay the biolayer. *Sensors Actuators B*, **44**, 578–584.

Beyer, M., C. Menzel, R. Quack, T. Scheper, K. Schugerl, W. Treichel, H. Voigt, M. Urich and R. Ferretti (1994). Development and application of a new enzyme sensor type based on the EIS-capacitance structure for bioprocess control. *Biosensors Bioelectron.*, **9**, 17–21.

Bindra, D.S., Y. Zhang, G.S. Wilson, R. Sternberg, D.R. Thevenot, D. Moatti and G. Reach (1991). Design and in vitro studies of a needle type glucose sensor for subcutaneous monitoring. *Anal. Chem.*, **63**, 1692–1696.

Blackborn, G.F. (1987). Enzyme-based CHEMFET. In *Biosensors: Fundamentals and Applications*. A.P.F. Turner, I. Karube and G.S. Wilson, Oxford University Press, New York, pp. 481–530.

Blackburn, G.F., D.B. Talley, P.M. Booth, C.N. Durfor, M.T. Mertin, A.D. Napper and A.R. Rees (1990). Potentiometric biosensor employing catalytic antibodies as the molecular recognition element. *Anal. Chem.*, **62**, 2211–2216.

Bogdanovskaya, V.A., V.A. Fridman, M.R. Tarasevich and F. Scheller (1994). Bioelectrocatalysis by immobilised peroxidase: the reaction mechanism and the possibility of electrochemical detection of both inhibitors and activators of enzyme. *Anal. Lett.*, **27**, 2833–2847.

Bonakdar, M., J.L. Vilchez and H.A. Mottola (1989). Bioamperometric sensors for phenol based on carbon paste electrodes. *J. Electroanal. Chem.*, **266**, 47–55.

Brooks, G.T., and T.R. Roberts (Eds.) (1999). *Pesticide Chemistry and Bioscience*. Royal Society of Chemistry, London.

Buch, R.M., T.Q. Barker and G.A. Rechnitz (1991). Intact chemoreceptor biosensors based on Hawaiian aquatic species. *Anal. Chim. Acta*, **243**, 157–166.

Campanella, L., T. Beone, M.P. Sammartino and M. Tomasetti (1993). Determination of phenol in waste and water using an enzyme sensor. *Analyst*, **118**, 979–986.

Caras, S., and J. Janata (1980). Field-effect transistor sensitive to penicillin. *Anal. Chem.*, **52**, 1935–1937.

Caras, S., and J. Janata (1985). pH-based enzyme potentiometric sensors. *Anal. Chem.*, **57**, 1917–1925.

Cardosi, M.F., and A.P.F. Turner (1991). Mediated Electrochemistry: a practical approach to biosensing, In *Advances in Biosensors*, Vol. 1. A.P.F. Turner (Ed.), JAI Press, London, pp. 125–169.

Cass, A.E.G., G. Davis, G.D. Francis, H.A.O. Hill, W.J. Aston, I.J. Higgins, E.V. Plotkin, L.D.L. Scott and A.P.F. Turner (1984). Ferrocene-mediated enzyme electrode for amperometric detection of glucose. *Anal. Chem.*, **56**, 667–671.

Chan, W.H., and T.Y. Xie (1997). Adsorption voltammetric determination of μg levels formaldehyde via in situ derivatisation with Girard reagent T. *Anal. Chim. Acta*, **339**, 173–179.

Chaubey, A., and B.D. Malhotra (2002). *Biosensors Bioelectron.*, **17**, 441–456.

Chaubey, A., M. Gerard, R. Singhal, V.S. Singh and B.D. Malhotra (2000). Immobilization of lactate dehydrogenase on electrochemically prepared polypyrrole–polyvinyl sulphonate composite films for application to lactate biosensors. *Electrochim. Acta*, **46**, 723–729.

Chen, F., and L. Shenghui (2000). A review of study on DNA biosensor. II. Electrochemical DNA biosensor. *J. Zhejiang Univ. Technol.*, **28**(1), 47–50.

Chin, Q.J., and S.J. Dong (1994). Electrocatalytic oxidation of reduced nicotinamide coenzymes at methylene-green-modified electrodes and fabrication of amperometric alcohal biosensors. *Anal. Chim. Acta*, **285**, 125–133.

Clark, L.C., and C. Lyons (1962). Electrode systems for continuous monitoring in cardiovascular surgery. *Ann. N.Y. Acad Sci.*, **102**, 29–45.

Collier, W.A., D. Janssen and A.L. Hart (1996). Measurement of soluble L-lactate in dairy products using screen-printed sensors in batch mode. *Biosensors Bioelectron.*, **11**, 1041–1049.

Collier, W.A., P. Lovejoy and A.L. Hart (1998). Estimation of soluble L-lactate in dairy products using screen-printed sensors in a flow injection analyzer. *Biosensors Bioelectron.*, **13**, 219–225.

Colston, J.T., P. Kumar, E.D. Rael, A.T.O. Tsin, J.J. Valdes and J.P. Chambers (1993). Detection of sub-nanogram quantities of Mojave toxin via enzyme immunoassay with light-addressable potentiometric detector. *Biosensors Bioelectron.*, **8**, 117–121.

Cosnier, S. (1999). Biomolecule immobilization on electrode surfaces by entrapment or attachment to electrochemically polymerized films. A review. *Biosensors Bioelectron.*, **14**, 443–456.

Cosnier, S., and C. Innocent (1993). A new strategy for the construction of a tyrosinase-based amperometric phenol and o-diphenol sensor. *Biochem. Bioenerg.*, **31**, 147–160.

Cosnier, S., B. Galland, Ch. Gondranc and A. Le Pellec (1998). Electrogeneration of biotinylated functionalised polypyrrole for the simple immobilization of enzymes. *Electroanalysis*, **10**, 808–813.

Cosnier, S., J.-J. Fombon, P. Labbe and D. Limosin (1999). Development of a PPO-poly(amphiphilic pyrrole) electrode for on site monitoring of phenol in aqueous effluents. *Sensors Actuators B*, B59, 134–139.

Daily, S., S.J. Armfield, B.G.D. Haggett and M.E.A. Downs (1991). Automated enzyme packed-bed system for the determination of vitamin C in food stuffs. *Analyst*, **116**, 569–572.

Danilowicz, C., E. Corton, F. Battaglini and E.J. Calvo (1999). An $Os(byp)_2Cl–PyCH_2NH$ poly(allylamine) hydrogel mediator for enzyme wiring at electrodes. *Electrochim. Acta*, **43**, 3525–3531.

Dave, K.I., C.E. Miller and J.R. Wild (1993). Characterization of organophosphorous hydrolyses and genetic manipulation of the phosphotriesterase from pseudomonas diminuta. *Chem. Biol. Interact.*, **87**, 55–68.

Degani, Y., and A. Heller (1989). Electrical communication between redox centers of glucose oxidase and electrodes via electrostatically and covalently bound redox polymers. *J. Am. Chem. Soc.*, **111**, 2357–2358.

Dempsey, E., O. Diamond, M.R. Smyth, G. Urban, G. Jobst, I. Moser, E.M.J. Verpoorte, A. Manz, H.M. Widmer, K. Robenstein and R. Freaney (1997). Design and development of a miniaturized total chemical analysis system for on-line lactate and glucose monitoring in biological samples. *Anal. Chim. Acta*, **346**, 341–349.

Dennison, M.J., J.M. Hall and A.P.F. Turner (1995). Gas-phase microbiosensor for monitoring phenol vapor at ppb levels. *Anal. Chem.*, **67**, 239–246.

Deshpande, M.N., and J. Amalnekar (1993). Biosensors preparation from electrochemically synthesized conducting polymers. *Prog. Polymer Sci.*, **18**, 623–649.

DeSilva, M.S., Y. Zhang, P.J. Hesketh, J. Maclay, S.M. Gendel and J.R. Stetter (1995). Impedance based sensing of the specific binding reaction between Staphylococcus entertoxin B and its antibody on an ultra-thin platinum film. *Biosensors Bioelectron.*, **10**, 675–682.

Dill, K., J.A. Blomdahl, S.R. Lyon and J.D. Olson (1994). Improvement in threshold immunoassay technology. In *Proceedings of 1993 ERDEC Scientific Conference on Chemical Defence Research*, 16–19 November. D.A. Berg, J.D. Williams Jr. and P.J. Reeves (Eds.). Report no. ERDEC-SP-024, August 1994, pp. 45–50.

Dransfeld, I., R. Hintsche, W. Moritz, M.T. Pham, W. Hoffmann and J. Hueller (1990). Biosensors for glucose and lactate using fluoride-ion sensitive field effect transistors. *Anal. Chem.*, **23**, 437–450.

Dumschat, C., H. Muller, K. Stein and G. Schwedt (1991). Pesticide sensitive ISFET based on enzyme inhibition. *Anal. Chim. Acta*, **252**, 7–9.

Dzyadevich, S.V., Y.I. Korpan, V.N. Arkhipova, M.Yu. Alesina, C. Martelet, A.V. El'skaya and A.P. Solatkin (1999). Application of enzyme field effect transistors for the detection of glucose concentration in blood serum. *Biosensors Bioelectron.*, **14**, 283–287.

Eftekhari, A. (2001). Aluminum as a suitable substrate for the deposition of conducting polymers: application to polyaniline and enzyme modified electrodes. *Synth. Met.*, **125**, 295–300.

Eggenstein, C., M. Borchardt, C. Diekmann, B. Grundig, C. Dumschat, K. Cammann, M. Knoll and F. Spener (1999). A disposable biosensor for urea determination in blood based on an ammonia-sensitive transducer. *Biosensors Bioelectron.*, **14**, 33–41.

Emneus, J., and L. Gorton (1990). Flow system for starch determination based on consecutive enzyme steps and amperometric detection at a chemically modified electrode. *Anal. Chem.*, **62**, 263–268.

Erickson, K.A., and P. Wilding (1993). Evaluation of a novel print-of-care system, the I-STAT portable clinical analyzer. *Clin. Chem.*, **39**(2), 287.

Erlenkotter, A., M. Kottubus and G.C. Chemnitius (2000). Flexible amperometric transducers for biosensors based on a screen-printed three electrode system. *J. Electroanal. Chem.*, **481**, 82–94.

Fare, T.L., M.D. Cabelli, S.M. Dallas and D.P. Herzog (1998). Functional characterization of a conducting polymer-based immunoassay system. *Biosensors Bioelectron.*, **13**(3–4), 459–470.

Faridnia, M.H., G. Palleschi, G.J. Lubrano and G.G. Guilbault (1993). Amperometric biosensor for determination of lactate in sweat. *Anal. Chim. Acta*, **278**, 35–40.

Fernandez-Alba, A.R., L.H. Guil, G.D. Lopez and Y. Chisti (2001). Toxicity of pesticides in wastewater: a comparative assessment of rapid bioassays. *Anal. Chim. Acta*, **426**, 289–301.

Fernandez-Sanchez, C., M.B. Gonzalez-Garcia and A. Costa-Garcia (2000). AC voltammetric carbon paste-based enzyme immunosensors. *Biosensors Bioelectron.*, **14**, 917–924.

Flounders, A.W., A.K. Singh, J.V. Volponi, S.C. Carichner, K. Wally, A.S. Simonian, J.R. Wild and J.S. Schoeniger (1999). Development of sensors for direct detection of organophosphates. Part II: Sol–gel modified field effect transistor with immobilized organophosphate hydrolase. *Biosensors Bioelectron.*, **14**, 715–722.

Foulds, N.C., and C.R. Lowe (1988). Immobilisation of glucose oxidase in ferocene-modified pyrrole polymers. *Anal. Chem.*, **60**, 2473–2478.

Fromherz, P., A. Offenhausser, T. Vetter and J. Weiss (1991). A neuron–silicon junction: a retzius cell of the leech on an insulated-gate field-effect transistor. *Science*, **252**, 1290–1293.

Fromherz, P., C.O. Muller and R. Weiss (1993). Neuron transistor: electron transfer function measured by the patch-clamp technique. *Phys. Rev. Lett.*, **71**, 4079–4082.

Fuji, M., Y. Saeki, K. Arri and K. Yoshino (1990). Fractal structure of electrochemically polymerized polypyrrole and growth process as function of monomer concentration, electrolytic concentration and applied voltage. *Jpn. J. Appl. Phys.*, **29**, 2501–2505.

Garjonyte, R., and A. Malinauskas (2000). Glucose biosensor based on glucose oxidase immobilized in electropolymerized polypyrrole and poly(o-phenylenediamine) films on a Prussian Blue-modified electrode. *Sensors Actuators B*, **63**, 122–128.

Gaspar, S., K. Habermuller, E. Csoregi and W. Schuhmann (2001). Hydrogen peroxide sensitive biosensor based on plant peroxidases entrapped in Os-modified polypyrrole films. *Sensors Actuators B*, **72**, 63–68.

Genesis report (1992). June/July, pp. 16–20.

Gibson, T.D., J.N. Hulbert, S.M. Parkar, J.R. Woodward and I.J. Higgins (1992). Extended shelf-life of enzyme-based biosensors using a novel stabilisation system. *Biosensors Bioelectron.*, **7**, 701–708.

Gilmartin, M.A.T., and J.P. Hart (1994). Development of one-shot biosensor for the measurement of uric acid and cholesterol. *Analyst*, **119**, 2331–2336.

Gobi, K.V., and F. Mizutani (2001). Layer-by-layer construction of an active multilayer enzyme electrode applicable for direct determination of cholesterol. *Sensors Actuators B*, **80**, 272–277.

Gorton, L. (1995). Carbon paste electrodes modified with enzymes, tissues and cells. *Electroanalysis*, **7**, 23–45.

Gorton, L., H.I. Karan, P.D. Hale, T. Inagaki, Y. Ukamoto and T.A. Skotheim (1990). A glucose electrode based on carbon paste chemically modified with ferrocene-containing siloxane polymer and glucose oxidase coated with a poly(ester sulphonic acid) cation exchanger. *Anal. Chim. Acta*, **225**, 23–30.

Greenway, R.M., S.J. Gardiner, A.L. Hart and J. Dunlop (1994). Stainless steel electrodes in enzyme based potentiometric and pH stat biosensors. *Biosensors Bioelectron.*, **9**, 457–461.

Gruss, R., and F. Scheller (1987). An electrochemical method for the determination of cholinesterase in serum. *Z. Med. Lab. Diagn.*, **28**, 333–334.

Guilbault, C.G. (1984). *Analytical Uses of Immobilized Enzymes*. Marcel Dekker, New York.

Guilbault, G.G., and P.R. Coulet (1983). Creatinine-selective enzyme electrodes. *Anal. Chim. Acta*, **152**, 223–228.

Guilbault, G.G., and G.L. Lubrano (1973). An enzyme electrode for amperometric determination of glucose and lactate. *Anal. Chim. Acta*, **64**, 439–445.

Guilbault, G.G., and J. Montalvo (1969). A urea specific enzyme electrode. *J. Am. Chem. Soc.*, **91**, 2164–2169.

Guilbault, G.G., and G. Palleschi (1995). Non-invasive biosensors in clinical analysis. *Biosensors Bioelectron.*, **10**, 379–392.

Habermuller, K., A. Ramanavicius, V. Laurinavicius and W. Schuhmann (2000). An oxygen insensitive reagentless glucose biosensor based on osmium-complex modified polypyrrole. *Electroanalysis*, **12**, 1383–1389.

Hafeman, D., J. Parce and H. McConnel (1988). Light-addressable potentiometric sensor for biochemical systems. *Science*, **240**, 1182–1185.

Hahizadeh, K., H.T. Tang, H.B. Halsall and W.R. Heinemann (1991). Chemical crosslinking of a redox mediator thionin for electrolytic oxidation of NADH-cationic styrene and ethylenimine polymers. *Electroanalysis*, **8**, 575–581.

Haiyan, W., and M. Shaolin (1999). Bioelectrochemical characteristics of cholesterol oxidase immobilized in a polyaniline film. *Sensors Actuators B*, **56**, 22–30.

Hall, E.A.H. (1990). *Biosensors*. Open University Press, Buckingham, UK, pp. 1–351.

Hall, E.A.H., M. Preuss, J.J. Gooding and M. Hammerle (1998). Exploring sensors to monitor some environmental discharges. In *Biosensors for Direct Monitoring of Environmental Pollutants in Field*, Vol. 38 of NATO ASI series 2 – Environment. D.P. Nikolelis, U.J. Krull, J. Wang and M. Mascini (Eds.), Kluwer, Dordrecht, pp. 227–237.

Hall, G.F., and A.P.F. Turner (1991). Organic-phase enzyme electrode for cholesterol. *Anal. Lett.*, **24**, 1375–1388.

Hall, J.M., J. Moore-Smith, J.V. Bannister and I.J. Higgins (1994). An electrochemical method for detection of nucleic acid hybridization. *Biochem. Mol. Bio. Int.*, **32**, 21–28.

Hammerle, M., E.A.H. Hall, N. Cade and D. Hodgins (1996). Electrochemical enzyme sensor for formaldehyde operating in the gas phase. *Biosensors Bioelectron.*, **11**, 239–246.

Hart, A.L., C. Matthews and W.A. Collier (1999). Estimation of lactate in meat extracts by screen-printed sensors. *Anal. Chim. Acta*, **386**, 7–12.

Hashimoto, K., K. Ito and Y. Ishimori (1998). Microfabricated disposable DNA sensor for detection of hepatitis B virus DNA. *Sensors Actuators B*, **46**, 220–225.

Hikuma, M., H. Suzuki, T. Yasuda, I. Karube and S. Suzuki (1979). Amperometric estimation of BOD by using living immmobilised yeasts. *Fur. J. Appl. Microb. Biotechnol.*, **8**, 289–297.

Ho, W.O., S. Krause, C.J. McNeil, J.A. Pritchart, R.D. Armstrong, D. Athey and K. Rawson (1999). Electrochemical sensor for measurement of urea and creatinine in serum based on impedance measurement of enzyme-catalysed polymer transformation. *Anal. Chem.*, 1940–1946.

Hu, X., and Z. Leng (1995). Determination of cyanide using a tryosinase amperometric biosensor with catechol as substrate. *Analyst*, **120**, 1555–1557.

Huan, H., B. Persson, L. Gorton, S. Sahni, T.A. Skotheim and P. Bartlett (1996). Redox polymers for electrocatalytic oxidation of NADH-cationic styrene and ethylenimine polymers. *Electroanalysis*, **8**, 575–581.

Ivnitski, D., I. Abdel-Hamid, P. Atanasov and E. Wilkins (1999). Biosensors for detection of pathogenic bacteria. *Biosensors Bioelectron.*, **14**, 599–624.

Jaegfeldt, H. (1980). Adsorption and electrochemical oxidation behaviour of NADH at a clean platinum electrode. *J. Electroanal. Chem.*, **110**, 295–302.

Jing, L., L. Bjornsson and B. Mattiasson (2000). Immobilized activated sludge based biosensor for biochemical oxygen demand measurement. *Biosensors Bioelectron.*, **14**, 883–893.

Jones, A.W. (1979). Assessment of an automated enzymatic method for ethanol determination in microsamples of saliva. *Scand. J. Clin. Lab. Invest.*, **39**, 199–203.

Kaisheva, A., I. Iliev, S. Christov and R. Kazareva (1997). Electrochemical gas biosensor for phenol. *Sensors Actuators*, **44**, 571–577.

Kaissling, K.E. (1995). Single unit and electroantennogram recordings in insect olfactory organs. In *Experimental Cell Biology of Taste and Olfaction*. A.I. Spielman and J.G. Brand (Eds.), CRC Press, Boca Raton, FL, p. 361.

Katakis, I., and E. Dominguez (1997). Catalytic electroxidation of NADH for dehydrogenase amperometric biosensors. *Microchim. Acta*, **126**, 11–32.

Keedy, F.H., and P. Vadgama (1991). Determination of urate in undiluted and whole blood for enzyme electrode. *Biosensors Bioelectron.*, **6**, 491–499.

Khan, G.F., and W. Werner (1997). A highly sensitive amperometric creatinine sensor. *Anal. Chim. Acta*, **351**, 151–158.

Kim, J.H., J.H. Cho, G.S. Cha, C.W. Lee, H.B. Kim and S.H. Paek (2000). Conductimetric membrane strip immunosensor with polyaniline-bound gold colloids as signal generator. *Biosensors Bioelectron.*, **14**, 907–915.

Koopal, C.G.J., A.A.C.M. Bos and R.J.M. Nolte (1994). Third-generation glucose biosensor incorporated in a conducting printing ink. *Sensors Actuators B*, **18**, 166–170.

Korpan, Y.I., M.V. Gonchar, N.F. Starodub, A.A. Shul'ga, A.A. Sibirny and A.V. El'skaya (1993). A cell biosensor specific for formaldehyde based on pH-sensitive transistors coupled to methylotrophic yeast cells with genetically-adjusted metabolism. *Anal. Biochem.*, **215**, 216–222.

Korpan, Y.I., A.P. Soldatkin, M.V. Gonchar, A.A. Sibirny, T.D. Gibson and A.V. El'skaya (1997). A novel enzyme biosensor specific for formaldehyde based on pH-sensitive field effect transistors. *J. Chem. Technol. Biotechnol.*, **68**, 209–213.

Korpan, Y.I., M.V. Gonchar, A.A. Sibirny, C. Martelet, A.V. El'skaya, T.D. Gibson and A.P. Soldatkin (2000). Development of highly selective and stable potentiometric sensors for formaldehyde determination. *Biosensors Bioelectron.*, **15**, 77–83.

Korri-Youssoufi, H., F. Garnier, P. Srivastava, P. Godillot and A. Yassar (1997). Toward bioelectronics: specific DNA recognition based on an oligonucleotide-functionalized polypyrrole. *J. Am. Chem. Soc.*, **119**, 7388–7389.

Kranz, C., H. Wohlschlager, H.-L. Schmidt and W. Schuhmann (1998). Controlled electrochemical preparation of amperometric biosensors based on conducting polymer multilayers. *Electroanalysis*, **10**, 546–552.

Kulys, J., and D. Schmid (1991). Bioenzyme sensors based on chemically modified electrodes. *Biosensors Bioelectron.*, **6**, 43–48.

Kumar, A., Rajesh, A. Chaubey, S.K. Grover and B.D. Malhotra (2001). Immobilization of cholesterol oxidase and potassium ferricyanide on dodecylbenzene sulfonate ion-doped polypyrrole film. *J. Appl. Polymer Sci.*, **82**, 3486–3491.

Lima Filho, J.L., P.C. Pandey and H.H. Weetall (1996). An amperometric flow injection analysis enzyme sensor for sucrose using a tetracyanoquinodimethane modified graphite paste electrode. *Biosensors Bioelectron.*, **11**, 719–723.

Lin, Y., and G.G. Wallace (1994). Electropolymerization of pyrrole under hydrodynamic conditions – effect of solution additives. *Electrochim. Acta*, **39**, 1409–1413.

Liu, J., and J. Wang (1999). Remarkable thermostability of bioelectrodes based on enzymes immobilised within hydrophobic semi-solid matrices. *Biotechnol. Appl. Biochem.*, **30**, 177–183.

Liu, J., L. Bjornsson and B. Mattiasson (2000). Immobilised activated sludge based biosensor for biochemical oxygen demand measurement. *Biosensors Bioelectron.*, **14**, 883–893.

Livache, T., A. Roget, E. Dejean, C. Barthet, G. Bidan and R. Teoule (1994). Biosensing effects in functionalized electroconducting conjugated polymer layers: addressable DNA matrix for the detection of gene mutations. *Synth. Met.*, **71**, 2143–2146.

Lobo, M.J., A.J. Miranda and P. Tunon (1997). Amperometric biosensors based on NAD(P)-dependent dehydrogenase enzymes. *Electroanalysis*, **9**, 191–202.

Lund, V., R. Schmid, D. Rickwood and E. Hornes (1988). Assessment of methods for covalent binding of nucleic acid to magnetic beads, DYNABEADS™, and the characteristics of the bond nucleic acids in hybridization reactions. *Nucleic Acids Res.*, **22**, 10861–10880.

Luth, H., M. Thust, A. Steffen, P. Kordos and M.J. Schoning (2000). Biochemical sensors with structured and porous silicon capacitors. *Mater. Sci. Eng. B*, **69–70**, 104–108.

Macholan, L. (1990). Phenol-sensitive enzyme electrode with substrate cycling for quantification of certain aromatic acids and thio compounds. *Collect. Czech. Chem. Commun.*, **55**, 2152–2160.

Macholan, L., and L. Schanel (1977). Enzyme electrode with immobilized polyphenol oxidase for the determination of phenolic substrates. *Collect. Czech. Chem. Commun.*, **42**, 3667–3675.

Madaras, M.B., and R.P. Buck (1996). Miniaturized biosensor employing electropolymerized permselective films and their use for creatinine assay in human serum. *Anal. Chem.*, **68**, 3832–3839.

Marraza, G., L. Chianella and M. Mascini (1999a). Disposable DNA electrochemical sensor for hybridization detection. *Biosensors Bioelectron.*, **14**, 43–51.

Marraza, G., I. Chianella and M. Mascini (1999b). Disposable DNA electrochemical biosensor for environmental monitoring. *Anal. Chim. Acta*, **387**, 297–307.

Marraza, G., G. Chiti, M. Mascini and M. Anichini (2000). Detection of human apolipoprotein E, genotypes by DNA electrochemical biosensor coupled with PCR. *Clin. Chem.*, **46**, 31–37.

Maruyama, K., J. Motonaka, Y. Mishima, Y. Matsuzaki, I. Nakabayashi and Y. Nakabayashi (2001). Detection of target DNA by electrochemical method. *Sensors Actuators B*, **76**, 215–219.

Mascini, M., S. Fortunati, D. Moscone and D. Palleschi (1985). Ammonia abatement in an enzymatic flow system for the determination of creatinine in blood sera and urine. *Anal. Chim. Acta*, **171**, 175–184.

Mascini, M., F. Mazzei, D. Moscone, G. Calabrese and M. Massi-Bendetti (1987). Lactate and pyruvate electrochemical biosensors for whole blood in extracorporeal experiments with an endocrine artificial pancrease. *Clin. Chem.*, **33**, 591–593.

McGall, G.H., A.D. Barone, M. Diggelmann, S.P.A. Foder, E. Gentalen and N. Ngo (1987). The efficiency of light-directed synthesis of DNA arrays on glass substrates. *J. Am. Chem. Soc.*, **119**, 5081–5090.

Menzel, C., T. Lerch, T. Scheper and K. Schugerl (1995). Development of biosensors based on an electrolyte isolator semiconductor (EIS) capacitor structure and their application for process monitoring. *Anal. Chim. Acta*, **317**, 259–264.

Merkoci, A., E. Fabregas and S. Alegret (1999). Consolidated biocomposite membrane technology for production of potentiometric biosensors. *Sensors Actuators B*, **60**, 97–105.

Metzger, J., M. Reiss and W. Hartmeier (1998). Amperometric phenol biosensor based on a thermostable phenol hydroxylase. *Biosensors Bioelectron.*, **13**, 1077–1082.

Meyerhoft, M., and G.A. Rechnitz (1976). An activated enzyme electrode for creatinine. *Anal. Chim. Acta*, **85**, 277–285.

Mikkelsen, S.K. (1996). Electrochemical biosensors for DNA sequence detection. *Electroanalysis*, **8**, 15–19.

Mikkelsen, S.R., and G.A. Rechnitz (1989). Conductometric transducers for enzyme-based biosensors. *Anal. Chem.*, **61**, 1737–1742.

Millan, K.M., and S.R. Mikkelsen (1993). Sequence-selective biosensor for DNA based on electroactive hybridization indicators. *Anal. Chem.*, **65**, 2317–2323.

Millan, K.M., A. Sarullo and S.R. Mikkelsen (1994). Voltammetric DNA biosensor for cystic fibrosis based on a modified carbon paste electrode. *Anal. Chem.*, **66**, 2943–2948.

Mitzutani, F., S. Xabuki, Y. Sato and Y. Hirata (1998). Amperometric biosensors using an enzyme-containing polyion complex. *Polymer. Sens.*, **690**, 45–56.

Miyahara, Y., T. Marizumi and K. Ichimura (1985). Integrated enzyme FETs for simulataneous detection of urea and glucose. *Sensors Actuators B*, **7**, 1–10.

Moiroux, J., and P. Elving (1978). Effects of adsorption, electrode material, and operational variables on the oxidation of dihydronicotinamide adenin dinucleotide at carbon electrodes. *Anal. Chem.*, **50**, 1056–1062.

Moser, I., T. Schalkhammer, F. Pittner and G. Urban (1997). Surface techniques for electrochemical DNA biosensor. *Biosensors Bioelectron.*, **12**, 729–737.

Motonaka, J., and L.R. Faulkner (1993). Determination of cholesterol and cholesterol ester with novel enzyme micro-sensor. *Anal. Chem.*, **65**, 3258–3261.

Mullis, K., and F.A. Faloona (1987). Specific synthesis of DNA *in vitro via* a polymerase-catalyzed chain reaction. *Methods Enzymol.*, **155**, 335–350.

Mullor, S.G., M. Sanchez-Cabezudo, A.J.M. Ordieres and B.L. Ruiz (1996). Alcohol biosensor based on alcohol dehydrogenase and meldola blue immobilized into a carbon paste electrode. *Talanta*, **43**, 779–784.

Munoz, J., C. Jimenez, A. Bratov, J. Bartroli, S. Alegret and C. Dominguez (1997). Photosensitive polyurethanes applied to the development of CHEMFET and ENFET devices for biomedical sensing. *Biosensors Bioelectron.*, **12**, 577–585.

Nakaminami, J., S. Ito, S. Kuwabata and H. Yoneyama (1999). A biomimetic phospholipids/alkanethiolate bilayer immobilizing uricase and an electron mediator on an Au electrode for amperometric determination of uric acid. *Anal Chem.*, **71**, 4278–4283.

Nakaminami, T., S. Kuwabata and H. Yoneyama (1997). Electrochemical oxidation of cholesterol catalyzed by cholesterol oxidase with use of an artificial electron mediator. *Anal. Chem.*, **69**, 2367–2372.

Nanjo, M., and G.G. Guilbault (1975). Amperometric determination of alcohols, aldehydes and carboxylic acids with an immobilized alcohol oxidase enzyme electrode. *Anal. Chim. Acta*, **75**, 169–180.

Nguyen, A.L., and J.H.T. Luong (1993). Development of mediated amperometric biosensor for hypoxanthine glucose and lactate: a new formate. *Biosensors Bioelectron.*, **8**, 42–431.

Nikiforov, T.T., and Y.H. Rogers (1995). The use of 96-well polystyrene plates for DNA hybridisation based assays: an evaluation of different approaches to oligonucleotide immobilisation. *Anal. Biochem.*, **227**, 201–209.

Niu, J., and J.Y. Lee (2000). Bulk-modified amperometric biosensors for hypoxanthine based on sol–gel technique. *Sensors Actuators B*, **62**, 190–198.

Nowall, W.B., D.O. Wipf and W.G. Kuhr (1998). Localized avidin/biotin derivatization of glassy carbon electrodes using SECM. *Anal. Chem.*, **70**, 2601–2606.

Ohara, T.J., R. Royagopalan and A. Heller (1994). Wired enzyme electrodes for amperometric determination of glucose or lactate in the presence of interfering substance. *Anal. Chem.*, **66**, 2415–2475.

Onnerfjord, P., J. Emneus, G. Marko-Varga and L. Gorton (1995). Tyrosinase graphite-epoxy based composite electrodes for the detection of phenols. *Biosensors Bioelectron.*, 607–619.

Ortega, F., E. Dominguez, G. Jonsson-Pettersson and L. Gorton (1993). Amperometric biosensor for the determination of phenolic compounds using a tyrosinase graphite electrode in a flow injection system. *J. Biotechnol.*, **31**, 289–300.

Osaka, T., S. Komaba and A. Amano (1998). Highly sensitive microbiosensor for creatinine base on the combination of inactive polypyrrole with polyion complexes. *J. Electrochem. Soc.*, **145**, 406–408.

O'Sullivan, C.K., S. Lafis, M.I. Karayannis, M. Dennison, J.M. Hall, A.P.F. Turner, B. Hobbs and W.J. Aston (1996). The development of microsensors to monitor hazardous gases in the environment. In *Proceedings 4th European Workshop on Biosensors for Environmental Monitoring*, Barcelona, Spain, 15–16 February, pp. 44–46.

Owicki, J., L. Bousse, D. Hafeman, G. Kirk, J. Olson, H. Wada and J. Parce (1994). The light-addressable potentiometric sensor: principles and biological applications. *Annu. Rev. Biophys. Biomater. Struct.*, **23**, 87–113.

Pak, S., P.J. Hesketh, B. Ilic and W. Penrose (1998). Optimization of immobilization layer on impedance based immunobiosensor. In *Proceedings of the Symposium on Microstructures and Microfabricated Systems IV*, 1–6 November, Boston, MA, pp. 176–184.

Palanti, S., G. Marrazza and M. Mascini (1996). Electrochemical DNA Probes. *Anal. Lett.*, **29**, 2309–2331.

Palchetti, I., A. Cagnini, M. Mascini and A.P.F. Turner (1999). Characterisation of screen-printed electrodes for detection of heavy metals. *Microchim. Acta*, **131**, 65–73.

Pallechi, G., M.H. Fardinia, G.J. Lubrano and G.G. Guilbault (1991). Ideal hydrogen peroxide-based glucose sensor. *Appl. Biochem. Biotech.*, **31**, 21–35.

Palleschi, G., and A.P.F. Turner (1990). Amperometric tetrafulvalene-mediated lactate electrode using lactate oxidase absorbed on carbon foil. *Anal. Chim. Acta*, **234**, 1459–496.

Palmisano, F., D. Centonize and P.G. Zambonin (1994). An in situ electrosynthesised amperometric biosensor based on lactate oxidase, immobilized in a poly o-phenylene-diamine film: determination of lactate in serum by flow injection analyzer. *Biosensors Bioelectron.*, **9**, 471–479.

Palmqvist, E., M. Khayyami, B. Danielsson, P.O. Larsson, K. Mosbach, C. Berggren and D. Kriz (1994). Development of a simple detector for microbial metabolism, based on a polypyrrole DC resistometric device. *Biosensors Bioelectron.*, 9. 551.

Palmqvist, E., C.B. Kriz, K. Svanberg, M. Khayyami and D. Kriz (1995). DC resistometric urea sensitive device using a conducting polymer film for the gas-phase detection of ammonia. *Biosensors Bioelectron.*, **10**, 283–287.

Pandey, P.C. (1988). A new conducting polymer–coated glucose sensor. *J. Chem. Soc. Faraday Trans. I*, **84**, 2259–2265.

Pandey, P.C., and A.P. Mishra (1988). Conducting polymer-coated enzyme microsensor for urea. *Analyst*, **113**, 329–331.

Pandey, P.C., and S. Upadhyay (2001). Bioelectrochemistry of glucose oxidase immobilized on ferrocene encapsulated ormosil modified electrode. *Sensors Actuators B*, **76**, 193–198.

Pandey, P.C., V. Pandey and S. Mehta (1993). A glucose sensor based graphite paste modified with tetracyanoquinodimethane. *Anal. Biochem.*, **214**, 233–237.

Pandey, P.C., S. Upadhyay, H.C. Pathak, C.M.D. Pandey and I. Tiwari (2000). Acetylthiocholine/acetylcholine and thiocholine/choline electrochemical biosensors/sensors based on an organically modified sol–gel glass enzyme reactor and graphite paste electrode. *Sensors Actuators B*, **62**, 109–116.

Park, J.K., H.J. Yee and S.T. Kim (1995). Amperometric biosensor for determination of ethanol vapor. *Biosensors Bioelectron.*, **10**, 587–594.

Park, J.K., H.J. Yee, K.S. Lee, W.Y. Lee, M.C. Shin, T.H. Kim and S.R. Kim (1999). Determination of breath alcohol using a differential-type amperometric biosensor based on alcohol dehydrogenase. *Anal. Chim. Acta*, **390**, 83–91.

Patel, N.G., A. Erlenkotter, K. Cammann and G.C. Chemnitius (2000). Fabrication and characterization of disposable type lactate oxidase sensors for dairy products and clinical analysis. *Sensors Actuators B*, **67**, 134–141.

Patel, N.G., S. Meier, K. Cammann and G.C. Chemnitius (2001). Screen-printed biosensors using different alcohol oxidases. *Sensors Actuators B*, **75**, 101–110.

Peguin, S., P.R. Coulet and G. Bardeletti (1989). Pyruvate oxidase and oxaloacetate decarboxylase enzyme electrodes. Simultaneous determination of transaminases with a two-electrode-based analyser. *Anal. Chim. Acta*, **222**, 83–93.

Persson, B., and L. Gorton (1990). A comparative study of some 3,7-diaminophenoxazine derivatives and related compounds for electrocatalytical oxidation of NADH. *J. Electroanal. Chem.*, **292**, 115–138.

Persson, B., H.L. Lan, L. Gorton, Y. Okamoto, P.D. Hale, L.I. Boguslavsky and T.A. Skotheim (1990). Amperometric biosensor based on electrocatalytic regeneration of NAD^+ at redo-polymer-modified electrodes. *Biosensors Bioelectron.*, **8**, 81–88.

Pfeiffer, D., B. Moller, N. Klimes, J. Szeponik and S. Fischer (1997). Amperometric lactate oxidase catheter for real-time lactate monitoring based on thin film technology. *Biosensors Bioelectron.*, **12**, 539–550.

Pilloton, R., T.N. Nwosu and M. Mascini (1988). Amperometric determination of lactic acid. Applications on samples. *Anal. Lett.*, **21**, 727–740.

Poghossian, A., M.J. Schoning, P. Schroth, A. Simonis and H. Luth (2001). An ISFET-based penicillin sensor with high sensitivity, low detection limit and long lifetime. *Sensors Actuators*, **76**, 519–526.

Przybyt, M., and H. Sugier (1990a). Metal–metal oxide enzyme electrodes for determination of urea. *Anal. Chim. Acta*, **237**, 399–404.

Przybyt, M., and H. Sugier (1990b). Tungsten electrode for urea. *Anal. Chim. Acta*, **239**, 269–276.

Puig-Lleixa, C., C. Jimenez and J. Bartroli (2001). Acrylated polyurethane-photopolymeric membrane for amperometric glucose biosensor construction. *Sensors Actuators B*, **72**, 56–62.

Rainina, E.I., E.N. Efremenco, S.D. Varfolomeyev, A.L. Simonian and J.R. Wild (1996). The development of a new biosensor based on recombinant E. coli for the direct detection of organophosphorous neurotoxins. *Biosensors Bioelectron.*, **11**, 991–1000.

Ramanathan, K., M.K. Ram, B.D. Malhotra and A.S.N. Murthy (1995). Application of polyaniline-Langmuir–Blodgett films as a glucose biosensor. *Mater. Sci. Eng. C*, **3**, 159–163.

Ramanathan, K., M.K. Ram, M.M. Verghese and B.D. Malhotra (1996). Dielectric spectroscopic studies on polypyrrole glucose oxidase films. *J. Appl. Polymer Sci.*, **60**, 2309–2316.

Reiter, S., K. Habermuller and W. Schuhmann (2001). A reagentless glucose biosensor based on glucose oxidase entrapped into osmium-complex modified polypyrrole films. *Sensors Actuators B*, **79**, 150–156.

Renger, G.T., A.T. Rene, A. Hoekstra, K. Venema and J. Korf (2001). An ultrafiltration catheter for monitoring of venous lactate and glucose around myocardial ischemia. *Biosensors Bioelectron.*, **16**, 159–167.

Rogers, K.R., M. Foley, S. Alter, P. Koga and M. Eldefrawi (1991). Light-addressable potentiometric biosensor for the detection of anticholinesterase. Anal. Lett., **24**, 191–198.

Rogers, K.R., J.J. Valdes and M. Eldefrawi (1992). Detection of nicotinic receptor ligands with a light-addressable potentiometric sensor. *Anal. Biochem.*, **202**, 111–116.

Romani, A., M. Minunni, N. Mulinacci, P. Pinelli, F.F. Vincieri, M. Del Carlo and M. Mascini (2000). Comparison among differential pulse voltammetry, amperometric biosensor, and HPLC/DAD analysis for polyphenol determination. *J. Agri. Food Chem.*, **48**, 1197–203.

Rosario, S.A., G.S. Cha, M.E. Meyerhoft and M. Trojanowicz (1990). Biosensors in flow-injection analysis. *Anal. Chem.*, **62**, 2418–2424.

Sampath, S., and O. Lev (1996). Inert metal-modified, composite ceramic–carbon, amperometric biosensors: renewable, controlled reactive layer. *Anal. Chem.*, **68**, 2015–2021.

Scheller, F., and F. Schubert (1992). *Biosensors*. Elsevier, Amsterdam.

Schneider, J., B. Gründig, R. Renneberg, K. Cammann, M.D. Madaras, R.P. Buck and K.-D. Vorlop (1996). Hydrogel matrix for three enzyme entrapment in creatine/creatinine amperometric biosensing. *Anal. Chim. Acta*, **325**, 161–167.

Schoning, M.J., S. Schutz, P. Schroth, B. Weißbecker, A. Steffen, P. Kordos, H.E. Hummel and H. Luth (1998a). A bioFET on the basis of intact insect antennae. *Sensors Actuators B*, **47**, 235–238.

Schoning, M.J., M. Thust, M. Muller-Veggian, P. Kordos and H. Luth (1998b). A novel silicon-based sensor array with capacitive EIS structure. *Sensors Actuators B*, **47**, 225–230.

Schubert, F., S. Saini, A.P.F. Turner and F. Scheller (1992). Organic phase enzyme electrodes for the determination of hydrogen peroxide and phenol. *Sensors Actuators B*, **7**, 408–411.

Schuhmann, W., C. Kranz, J. Huber and E. Wohlschlager (1993). Conducting polymer based amperometric enzyme electrodes. Towards the development of miniaturized reagentless biosensors. *Synth. Met.*, **61**, 31–35.

Schutz, S., B. Weißbecker and H.E. Hummel (1996). Biosensors for volatiles released by damaged plants. *Biosensors Bioelectron.*, **11**, 427–433.

Schutz, S., B. Weißbecker, H.E. Hummel, A. Schwarz, M.J. Schoning, A. Reimer, P. Kordos and H. Luth (1997). Field effect transistor–insect antenna junction. *Naturwissenschaften*, **84**, 86–88.

Scott, A.O. (1998). *Biosensors for Food Analysis*. The Royal Society of Chemistry, Cambridge.

Seki, A., S. Ikeda, I. Kubo and I. Karube (1998). Biosensors based on light-addressable potentiometric sensors for urea, penicillin and glucose. *Anal. Chim. Acta*, **379,** 9–13.

Sergeyeva, T.A., N.V. Lavrik, A.E. Rachkov, Z.I. Kazantseva and A.V. El'skaya (1998). An approach to conductometric immunosensor based on phthalocyanine thin film. *Biosensors Bioelectron.*, **13**, 359–369.

Shichiri, M., N. Asakawa, Y. Yamasaki, R. Kawamori and H. Abe (1986). Telemetry glucose monitoring device with needle-type glucose sensor: a useful tool for blood glucose monitoring in diabetic individuals. *Diabetes Care*, **9**, 298–301.

Shin, J.H., S.Y. Yoon, I.J. Yoon, S.H. Choi, S.D. Lee, H. Nam and G.S. Cha (1998). Potentiometric biosensors using immobilized enzyme layer mixed with hydrophilic polyurethane. *Sensors Actuators B*, **50**, 19–26.

Shin, Y.-T., and H.-J. Huang (1999). A creatinine modified polyaniline electrode for creatinine analysis. *Anal. Chim. Acta*, **392**, 143–150.

Shiono, S., Y. Hanazato and M. Nakako (1986). Urea and glucose sensors based on ion sensitive field effect transistor with photolithographically patterned enzyme mebrane. *Anal. Sci.*, **2**, 517–521.

Shriver-Lake, L.C., and F.S. Ligler (1995). Field screening methods for hazardous wastes and toxic chemicals. In *VIP-47, Proceedings of an International Symposium*, Las Vegas, NV, Vol. 2. Air and Waste Management Association, Pittsburg, PA, p. 107.

Shulga, A.A., S.V. Dzyadevich, A.P. Soldatkin, S.V. Patskovsky, V.I. Strikha and A.V. El'skaya (1994). Thin film conductometric biosensor for glucose and urea determination. *Biosensors Bioelectron.*, **9**, 217–219.

Shulga, A.A., K. Netchiporou, A.K. Sandrovsky, A.A. Abalov, O.S. Frolov, Yu.G. Kononenko, H. Maupas and C. Martelet (1995). Operation of an ISFET with non-insulated substrate directly exposed to the solution. *Sensors Actuators*, **30**, 101–105.

Singh, A.K., A.W. Flounders, J.V. Volponi, C.S. Ashley, K. Wally, A.S. Simonian and J.S. Schoeniger (1999). Development of sensors for direct detection of organophosphates. Part I: Immobilization, characterization and stabilization of acetylcholinesterase and organophosphate hydrolase on silica supports. *Biosensors Bioelectron.*, **14**, 703–713.

Smit, M.H., and E.G. Cass (1990). Cyanide detection using a substrate-regenerating, peroxide-based biosensor. *Anal. Chem.*, **62**, 2429–2436.

Smit, M.H., and G.A. Rechnitz (1993). Toxin detection using tyrosinase-coupled oxygen electrode. *Anal. Chem.*, **65**, 380–385.

Soldatkin, A.P., A.V. El'skaya, A.A. Shulga, A.S. Jdanova, S.V. Dzyadevich, N. Jaffrezic-Renault, C. Martelet and P. Clechet (1994). Glucose sensitive conductometric biosensor with additional nafion membrane: reduction of influence of buffer capacity on the sensor response and extension of its dynamic range. *Anal. Chim. Acta*, **288**, 197–203.

Soldatkin, A.P., D.V. Gorchkov, C. Martelet and N. Jaffrezic-Renault (1997). Application of charged polymeric materials as additional permselective membranes for modulation of the working characteristics of penicillin-sensitive ENFETs. *Mater. Sci. Eng. C*, **5**, 35–40.

Sprules, S.D., I.C. Hartley, R. Wedge, J.P. Hart and R. Pittson (1996). A disposable reagentless screen-printed amperometric biosensor for the measurement of alcohol in beverages. *Anal. Chim. Acta*, **329**, 215–221.

Steel, A.R., and T.M. Herne (1998). Electrochemical quantification of DNA immobilized on gold. *Anal. Chem.*, **70**, 4670–4677.

Sun, X., P. He, S. Liu, L. Ye and Y. Fang (1998). Immobilisation of single-stranded deoxyribonucleic acid on gold electrode with self-assembled aminoethanethiol monolayer for DNA electrochemical sensor applications. *Talanta*, **47**, 487–495.

Svorc, J., S. Miertus and M. Stedamky (1995). Electrochemical biosensor and process for their preparation. WO-A-97/02359, Germany.

Taguchi, H., N. Ishihara, K. Okumura and Y. Shimabayashi (1990). Investigation of methods of enzyme immmobilisation on a coated wire electrode for the development of micro-sensor. *Anal. Chim. Acta*, **228**, 159–162.

Tang, L.X., Z.B. Koochaki and P. Vadgama (1990). Composite liquid membrane for enzyme electrode construction. *Anal. Chim. Acta*, **232**, 357–365.

Tarasevich, M.R., V.A. Bogdanovskaya and G.V. Zhutaeva (1993). Electrochemical biosensors. *Electrochemistry*, **29**, 1362–1367.

Tarlov, M.J., S. Semancik and K.G. Kreider (1990). Mechanistic and response studies of iridium oxide pH sensors. *Sensors Actuators B*, **1**, 293–297.

Tatsuma, T., and T. Watanabe (1991). Oxidase/peroxide bilayer-modified electrodes as sesnors for lactate, pyruvate, cholesterol and uric acid. *Anal. Chim. Acta*, **242**, 85–89.

Thust, M., P. Schroth, A. Topler, M.J. Schoning, M. Muller-Veggian, P. Kordos and H. Luth (1998). Improving the detection limit of a capacitive enzyme sensor by means of a diffusion barrier. In *Proceedings of Eurosensors XII*, Southampton, UK, 13–16 September, pp. 507–510.

Tobalina, F., F. Pariente, L. Hernandez, H.D. Abruna and E. Lorenzo (1998). Carbon felt composite electrodes and their use in electrochemically sensing: a biosensor based on alcohol. *Anal. Chim. Acta*, **358**, 15–25.

Tobalina, F., F. Pariente, L. Hernandez, H.D. Abruna and E. Lorenzo (1999). Integrated ethanol biosensors based on carbon paste electrodes modified with [Re(Phen-dione)-$(CO)_3Cl$] and [Fe(Phen-dione)]$(PF_6)_2$. *Anal. Chim. Acta*, **395**, 17–26.

Tombelli, S., and M. Mascini (1998). Electrochemical biosensor for biogenic amines. A comparison between different approaches. *Anal. Chim. Acta*, **358**, 277–284.

Torstesson, A., and L. Gorton (1981). Catalytical oxidation of NADH by surface-modified graphite electrodes. *J. Electroanal. Chem.*, **130**, 199–207.

Tran-Minh, C., P.C. Pandey and S. Kumaran (1990). Studies on acetylcholine sensor and its analytical application based on the inhibition of cholinesterase. *Biosensors Bioelectron.*, **5**, 461–471.

Trojanowicz, M., and M.L. Hitchman (1996). Determination of pesticides using electrochemical biosensors. *Trends Anal. Chem.*, **15**, 38–45.

Trojanowicz, M., and T. Krawczyk (1995). Electrochemical biosensors based on enzymes immobilized in electropolymerized films (review). *Microchim. Acta*, **121**, 167–181.

Tsuchida, T., and K. Yoda (1983). Multi enzyme membrane electrodes for determination of creatinine and creatine in sensor. *Clin. Chem.*, **29**, 51–55.

Turner, A.P.F., I. Karube and G.S. Wilson (Eds.) (1987). *Biosensors: Fundamentals and Applications*. Oxford University Press, Oxford.

Uithoven, K.A., J.C. Schmidt and M.E. Ballman (2000). Rapid identification of biological warfare agents using an instrument employing a light addressable potentiometric sensor and a flow-through immunofiltration-enzyme assay system. *Biosensors Bioelectron.*, **14**, 761–70.

Updike, S.J., and G.P. Hicks (1967). The enzyme electrode. *Nature*, **214**, 986–988.

Vadgama, P., J. Spoors, X. Tang and C. Battersby (1989). The needle glucose electrode: in vitro performance and optimization for implantation. *Biomed. Biochim. Acta*, **48**, 935–942.

Van der Schoot, B.H., and P. Bergveld (1987–1988). ISFET-based enzyme sensors. *Biosensors*, **3**, 161–186.

Varadi, M., and N. Adanyi (1994). Application of biosensor with amperometric detection for determining ethanol. *Analyst*, **119**, 1843–1847.

Vianello, F., A. Stefani, M.L. Di Paolo, A. Rigo, A. Lui, B. Margesin, M. Zen, M. Scarpa and G. Soncini (1996). Potentiometric detection of formaldehyde in air by an aldehyde dehydrogenase FET. *Sensors Actuators B*, **37**, 49–54.

Vidal, J.C., E. Garcia and J.R. Castillo (1999). In situ preparation of overoxidized PPy/oPPD bilayer biosensors for the determination of glucose and cholesterol in serum. *Sensors Actuators B*, **57**, 219–226.

Volotovsky, V.V., V.I. Soldatkin and A.V. Elskaya (1996). Glucose-sensitive ion-selective field effect transistor based biosensor with additional positively charged membrane. Dynamic range extension and reduction of buffer concentration influence on the sensor response. *Anal. Chim. Acta*, **322**, 77–81.

Wang, J., Y. Lin and L. Chen (1993). Organic-phase biosensor for monitoring phenol and hydrogen peroxide in pharmaceutical antibacterial products. *Analyst*, **118**, 277–287.

Wang, J., Q. Chen, M. Pedrero and J.M. Pingarron (1995). Screen-printed amperometric biosensors for glucose and alcohols based on ruthenium-dispersed carbon inks. *Anal. Chim. Acta*, **300**, 111–116.

Wang, J., E. Palecek, P.E. Nielsen, G. Rivas, X. Cai, H. Shiraishi, N. Dontha, D. Luo and P.A.M. Faraias (1996). Peptide nucleic acid probes for sequence-specific DNA biosensors. *J. Am. Chem. Soc.*, **118**, 7667–7670.

Wang, J., X. Cai, G. Rivas, H. Shiraishi and N. Dontha (1997). Nucleic-acid immobilisation, recognition and detection at chronopotentiometric DNA chips. *Biosensors Bioelectron.*, **12**, 587–599.

Watson, L.D., P. Maynard, D.C. Cullen, R.S. Sethi, J. Brettle and C.R. Lowe (1987). A microelectronic conductimetric biosensor. *Biosensors*, **3**, 101–115.

Wilson, G.S., and D.R. Thévenot (1989). Unmediated amperometric enzyme electrode. In *Biosensors: A Practical Approach*. T. Cass (Ed.), IRL Press, Oxford, pp. 1–15.

Winter, B., and K. Cammann (1989). Formaldehyde analysis by electrochemical biosensor – FIA. *Fres. Z. Anal. Chem.*, 334, 720.

Yamato, H., M. Ohawa and W. Wernet (1995). A polypyrrole/three enzyme electrode for creatinine detection. *Anal. Chem.*, **67**, 2776–2780.

Yao, T., and K. Takashima (1998). Amperometric biosensor with a composite membrane of sol–gel derived enzyme film and electrochemically generated poly(1,2-diaminobenzene) film. *Biosensors Bioelectron.*, **13**, 67–73.

Yon Hin, B.F.Y., and C.R. Lowe (1992). Amperometric response of polypyrrole entrapped bienzyme films. *Sensors Actuators B*, **7**, 339–342.

Yon Hin, B.F.Y., R.S. Sethi and C.R. Lowe (1990). Multi-analyte microeletronic biosensors. *Sensors Actuators*, **1**, 550–554.

Yuan, Y.J., H. Zhao, G.G. Wallace and S.B. Adeloju (2000). Effect of an intermediate on the amperometric response of a polypyrrole-based formate biosensing membrane. *Electrochem. Commun.*, **2**, 27–31.

Yulaev, M.F., R.A. Sitdykov, N.M. Dmitrieva, B.B. Dzantiev, A.V. Zherdev and K.A. Askarov (1995). Potentiometric immunoenzyme electrode for 2,4-dichlorophenoxyacetic acid. *J. Anal. Chem.*, **50**, 194–197.

Yulaev, M.F., R.A. Sitdikov, N.M. Dmitrieva, E.V. Yaznina, A.V. Zherdev and B.B. Dzantiev (2001). Development of a potentiometric immunosensor for herbicide simazine and its application for food testing. *Sensors Actuators B*, **75**, 129–135.

Zolfino, I., G. Calderaro, G. Del Carlo and L. Piras (1997). A microorganism based biosensor for the detection of urea by Proteus vulgaris. In *Proceedings of the 2nd Italian Conference on Sensors and Microsystems. Artificial and Natural Perception*, Rome, Italy, 3–5 February, pp. 99–103.

Chapter 4

DIAGNOSTICS APPLICATIONS OF ENZYME-DOPED SOL–GEL DERIVED GLASSES

Arun Kumar, M.N. Kamalasanan, Mangu Singh, Pratima Chauhan and B.D. Malhotra

OUTLINE

Advances in Biosensors
Volume 5, pages 101–130.

1. INTRODUCTION

In recent years, sol–gel derived materials have been gradually gaining importance for the fabrication of biosensors (Dave *et al.* 1994, Pankrotov and Lev 1995). The sol–gel technique is particularly advantageous for these applications because of a number of desirable features it offers such as room-temperature processability, control of surface properties, ability to form films/monoliths, and chemical inertness to biomolecules (Avnir 1995, Petit-Dominguez *et al.* 1997). Porous inorganic xerogels are particularly attractive matrices for electrochemical biosensors. The immobilized proteins are also found to be stable in the sol–gel environment. The sol–gel technique is advantageous since little or no heating is required in the preparation of the samples. To date, silicon alkoxide precursors have been studied the most extensively for the entrapment of biomolecules because they are inexpensive and exhibit relatively slow overall reaction kinetics. Thus, one can readily prepare silica sol–gels that are doped with a wide variety of reagents (e.g., chemical recognition elements) and tune the characteristics of the final glass matrix by adjusting the processing conditions (pH, precursor ratios etc.). The sol–gel processed materials have also been used for the development of ceramic films for conductive, mechanical and electro-optic applications (Avellaneda *et al.* 1994).

Apart from their optical and electronic applications, sol–gels have very interesting microstructural properties like large pore volume, pore size, surface area and grain size. The relative rate of hydrolysis and condensation as well as the size and extent of branching of the gel in the precursor solution prior to film deposition controls the sol–gel product. The controllable porosity attainable in the sol–gel process makes them a useful support for immobilization of large molecules like enzymes and provides a certain selectivity by restricting molecules which are to be excluded by preventing analyte species of larger size than the pores from entering the pores. They also have the advantage of containing a large number of interaction sites due to the large surface area of sol–gel films. Further, the film chemistry can be tailored so that they can bind certain species more strongly than others which are not to be detected (Cauqui and Rodriguez-Izquierdo 1992). These ceramic films are very resistant to chemicals and can withstand a large number of chemically reactive systems. Further, most enzymes encapsulated in sol–gel matrices give optically transparent glasses, which help in easier use of spectrophotometric techniques to quantitatively estimate the concentration of the reaction products (Livage 1996). The fine pore network in dried gels ($<10\,nm$) does not scatter visible radiation and allows diffusion of small molecules onto the electrode surface.

Presently researchers are using physical entrapment of molecules using the sol–gel process with a sandwich configuration for immobilization of the enzymes, which gives better results than the previous techniques.

2. PREPARATION OF DOPED SOL–GEL GLASSES

The sol–gel process comprises of hydrolysis and polycondensation of starting monomers to produce a colloidal suspension, the sol; the polymerization (gelation) of this wet network of porous metal oxide upon drying forms the xerogel or the dry gel. One of the technologically important aspects of sol–gel processing is that prior to gelation, the sol is ideal for preparing thin films by dipping or spinning (Hench and West 1990).

The polycondensation of the alkoxysilane is associated with gelation of the gel which after drying is densified by a mild heat treatment to form a glass. The properties of the final glass are determined by the chemical and physical conditions (Zhishang *et al.* 1993). During preparation, these depend upon the ratio of silane, alcohol, water, pH of the alkoxide, the presence of a catalyst, the drying time, and the amount of organic additives such as surface active agents. Pore size and surface area can be controlled by the addition of acid and base; for example, addition of NaF to the starting (TMOS) solution leads to increase in the average pore size (Matijevic 1988).

Three approaches are used for making sol–gel monoliths (Figure 1):

(i) gelation of a solution of colloidal powder;
(ii) hydrolysis and polycondensation of alkoxide followed by hydrolysis by hypercritical drying;

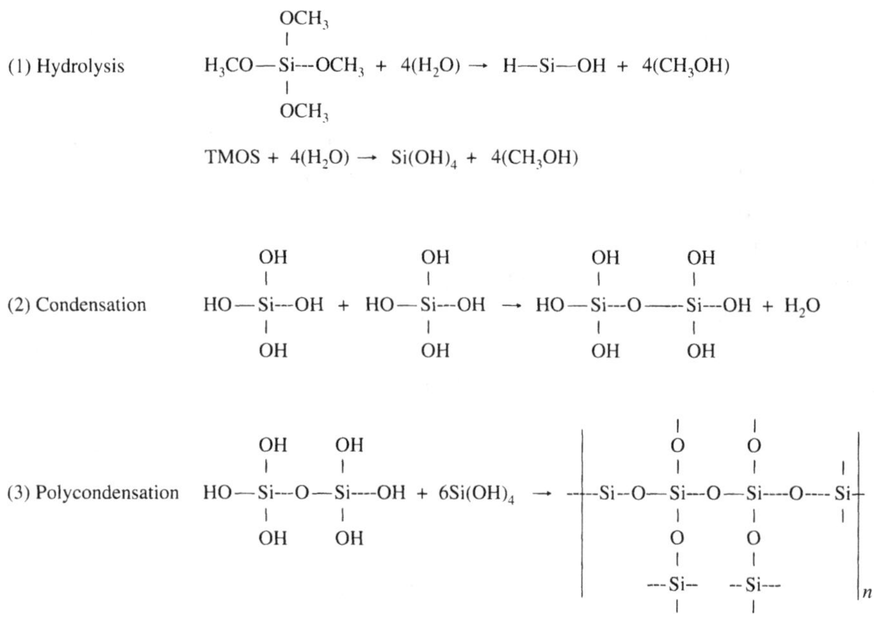

Figure 1. Steps involved in the preparation of sol–gel monoliths.

(iii) hydrolysis and polycondensation of alkoxide precursor followed by aging and drying under ambient atmosphere (Hardy *et al.* 1988).

The basic reactions involved in the preparation of sol–gel film are:

$$\text{Hydrolysis :} \quad \text{M–OR} + H_2O \rightarrow \text{M–OH} + \text{ROH}, \tag{1}$$

$$\text{Condensation :} \quad \text{M–OH} + \text{RO–M} \rightarrow \text{M–O–M} + \text{ROH}. \tag{2}$$

where $M = Si(OR)_4$ and $R = CH_3$, C_2H_5, or C_3H_7.

(i) Mixing: In this step a suspension of colloidal powder or sol is formed by mechanical mixing of colloidal particles. Alternately, a liquid alkoxide precursor such as $Si(OR)_4$ (where R is CH_3 or C_3H_7) is hydrolysed by mixing with water. The hydrated silica tetrahedra interact in a condensation reaction, forming Si–O–Si bonds. Linkage of additional Si–OH tetrahedra occurs as a polycondensation reaction and eventually results in a SiO_2 network. The hydrolysis and polycondensation reaction starts at various sites within the TMOS and H_2O solution as mixing occurs. When sufficient interconnected Si–O–Si bonds are formed in a region they are called colloidal (sub-micrometer) particles or a sol. The size of the sol particles and cross-linkage within the particles (i.e density) depends upon the pH and the ratio $R = [H_2O]/[Si(OR)_4]$, among other variables (Hench and Ulrich 1984).

(ii) Casting: The sol is a low-viscosity liquid. Therefore it can be cast into a mould.

(iii) Gelation: With time the colloidal particles and condensed silica species link together to form a three-dimensional network. The physical characteristics of the gel network depend upon the size of particles and the extent of cross-linkage prior to gelation. At gelation the viscosity increases sharply and a solid object results in the shape of the mould with appropriate control of the time-dependent change of viscosity of sol. Fibre can be pulled or film can be cast as gelation occurs (Klein 1988).

(iv) Aging: Aging of a gel, also called synergesis, maintains the cast object for long periods of time, hours to days, completely immersed in liquid. During aging polycondensation is continuous along with localized solution and re-precipitation of the gel network that increases the thickness of inter-particle necks, which results in decreased porosity. The strength of the gel, thereby, increases with aging. An aged gel develops sufficient strength to resist cracking during drying (Klein 1988).

(v) Drying: During drying the liquid is removed from the interconnected pore network. Large capillary stresses can develop during drying when the pores are small (<20 nm). These stresses will cause the gels to crack catastrophically unless the drying process is controlled by decreasing the liquid surface energy by (i) addition of surfactants or elimination of very small pores, (ii) hypercritical evaporation, which avoids the solid–liquid interface, or (iii) obtaining mono-dispersed pore size by controlling the rates of hydrolysis and condensation. After hypercritical drying the aerogel has a very low density and is a very good thermal insulator when sandwiched between evacuated glass plates.

(vi) Dehydration or Chemical Stabilization: The removal of surface silanol (Si–OH) bonds from the pore network results in a chemically stable ultraporous solid. Porous gel-silica made in this manner is optically transparent with interconnected porosity and has sufficient strength (Fegley and Barringer 1984).

(vii) Densification: Heating the porous gel to high temperatures causes densification: the pores are eliminated, and the density ultimately becomes equivalent to fused quartz or fused silica. The densification temperature depends considerably on the dimension of the pore network, the connectivity of the pores, and the surface area. Alkoxide gels are dried at low temperature.

2.1. Spin Coating Technique for Sol–Gel

The other methods adopted for the fabrication of an enzyme electrode with sol–gel techniques are spin coating and the sandwich configuration. Spin coating methods are used to cast the sol–gel material with enzyme on the desired substrate. Two types of substrates are used: either conducting or non-conducting. One takes an appropriate substrate and spin-coats onto this substrate a three-layer sol–gel coating followed by enzyme solution adsorption. It is then allowed to dry overnight at room temperature and another layer of sol–gel is cast at 3000 rpm for 30 s.

2.2. Sol–Gel Encapsulation Technique

In the sol–gel encapsulation method, enzyme solution in non-buffered water which may contain various additives is mixed in the cold (4°C) with either methanol or polyethylene glycol (PEG 400). The concentration of additives such as NaOH, HCl, etc. is controlled according to requirement. Tetramethyl orthosilicate (TMOS: 1 ml) is then added. The tubes containing the reaction mixture are transferred to a shaking water bath at 8°C. The bath is allowed to reach room temperature over 2–3 hours and during this period the enzyme (glucose oxidase) is added (Brinker and Scheterer 1990).

3. ADVANTAGES OF ENTRAPMENT OF PROTEINS IN SOL–GEL GLASSES

Biological macromolecules are highly efficient at recognizing specific analytes or catalyzing reaction in aqueous biological media. However an aqueous medium is generally necessary for biomolecular reactions and this limits their commercial viability. This is because any drastic changes in the aqueous medium often lead to partial or total denaturation and loss of catalytic activity of biomolecules. Efforts are thus being made to utilize these reagents in biosensors by immobilizing them in alternative environments that stabilize them and preserve their activity. Currently biosensors are being used in clinical testing and in industrial process monitoring. A biosensor consists of an immobilized biomolecule that reacts with a specific analyte, coupled with some type of signal transducer. One of the important areas of biosensor research is the immobilization of enzymes or proteins in an alternative environment that stabilizes them and preserves their reactivity (Livage 1996).

In the past, glasses and ceramics have been made only via solid-state reactions at temperatures above 1000°C. Therefore it was not possible to encapsulate any sensitive biological species within these matrices. Soft chemistry routes have been developed during the past two decades. They are performed under ambient conditions and open up new possibilities for materials chemistry. A sol is a dispersion of colloidal particles in a liquid. A gel is an interconnected rigid network with pores of sub-micrometer dimensions and polymeric chains whose average length is greater than a micrometer. The first sol–gel synthesis of silica was described about 150 years ago. Sol–gel chemistry is performed in solution at room temperature. Organic and inorganic precursors are mixed together and after polymerization, they form hybrid nanocomposites. Enzyme and antibody entrapment by sol–gel processing has received increased attention in recent years. Encapsulation of protein in transparent silicate glasses has been described (Wu *et al.* 1993). Silica biogels contain sufficient amounts of trapped interstitial water and provide an environment that is still aqueous, thereby contributing to the retention of structure and reactivity. Many researchers have used various techniques for immobilizing organic, inorganic and biological components. The bio-recognition property of the encapsulated proteins is unaltered by immobilization within the sol–gel matrix (Ellerby *et al.* 1992).

The major advantages of protein immobilization (Guilbault 1984) in sol–gel matrix are (i) close control of the reaction medium and conditions, (ii) prevention of bacterial and chemical degradation, (iii) cost-effective reusability of the protein, and (iv) enhanced biomolecular stability. Recent research has demonstrated that silicate glasses obtained by sol–gel methods can provide a host matrix, and the biomolecules immobilized on such matrices largely retain their functional characteristics and provide morphological and structural control that is not available when the biological molecules are simply dissolved in aqueous medium (Dave *et al.* 1994). For an optimum bio-stability and reaction efficiency, the preferred host matrix should isolate the biomolecule, protect it from self-aggregation and microbial attack, while providing the same local aqueous microenvironment as in biological media.

Recent sol–gel techniques enable bioactive composite layers to be prepared by the embedding of bioactive compounds, biomolecules and cellular systems within inorganic layers. These novel bioactive layers offer interesting new applications such as biocompatible coatings on implants and medical products, the preparation of biosensors and biocatalysts, and coatings that can release biocides in a controlled manner.

4. APPLICATION OF SOL–GEL GLASSES IN BIOSENSORS

The sol–gel process has been extensively used in the past decade for making large-area optical coatings, especially automobile rearview mirrors and high-reflectivity laser mirrors (Schubert *et al.* 1995). Immobilization coupled with retention of affinity or activity of biomolecules within the matrix have made sol–gels a potential vehicle for the development of new chemical biosensors (Dave *et al.* 1994) and optoelectronic sensors (Ulatowska *et al.* 2001). Properties of the sol–gel matrix like physical rigidity, negligible swelling in aqueous solutions, chemical inertness and thermal stability are important for biosensor development. The large size and high molecular weights of proteins prevent leaching from the sol–gel matrix and provide an efficient biosensing design in which the movement of the recognition molecule is restricted but the flow of smaller analytes through the pores of the gel is allowed. The gel entrapment is largely independent of the size of the biomolecule because the matrix is formed around it. In aqueous medium the side chain of a biomolecule interacts with the solvent (water) through hydrogen bonding and dipolar interactions. The recognition centres of a biomolecule are exposed on the surface and its efficiency is usually maximized in an aqueous medium (Brinker and Scheterer 1990).

The first reports of protein encapsulation inside silica glasses are relatively recent (Braun *et al.* 1992). They described the preparation of biochemically active glasses with the enzyme entrapped in the matrix. The resulting glasses retained 30% of the enzyme activity. Proteins other than enzymes have also been entrapped within sol–gel matrices. It has been shown that proteins such as myoglobin encapsulated in silica glasses are still able to react reversibly with O_2 or CO. Antibodies, polyclonal antifluorescein, have been encapsulated within a sol–gel (J. Wang 1999) and it was shown that fluorescein could diffuse into the porous silica matrix and bind selectively to the encapsulated antifluorescein (R. Wang *et al.* 1993). Tetraethyl orthosilicate-derived sol–gel film doped with glucose oxidase was used as a sensing platform for glucose (Narang *et al.* 1994a). Sol–gel organic–inorganic hybrid material based on silica sol and PVA-g-P(4-VP) copolymer has been used as a matrix for enzyme immobilization (B. Wang *et al.* 2000). The immobilization of antigen or antibody (Starodub *et al.* 1996) on porous silicon has been reported for the design of opto-electronic elements. The fluorescence intensity did not change significantly after the immobilization of antigen or antibody on the porous silicon surface. A highly stable silicon-based pH sensor (Schoning *et al.* 1996) utilizing a capacitive electrolyte insulator–semiconductor heterostructure was developed using a pulsed laser deposition technique. In another approach, silica and polyamide coating was successfully applied (Sun *et al.* 1996) on both sides of 10 MHz AT-Cu^+ quartz crystal resonators with silver and gold electrodes. This study highlighted that multisensors with various coatings and electrodes provide possibilities to monitor wastewater in real-time.

Three-dimensional structures including membranes, microvalves and pumps, channels, etc., in silicon have been developed. When integrated with microcircuits they provide the opportunity of creating unique miniature smart chemical sensing device (Tuller and Mlcuk 1996). Bacteriorhodopsin encapsulated (Wu *et al.* 1993) along with membrane in a sol–gel film has been utilized for the fabrication of an opto-electronic device. Dissolved oxygen has been measured through the oxygenation of deoxymyoglobin trapped in a sol–gel matrix (McCurley *et al.* 1996). The encapsulated myoglobin is reduced from met-myoglobin to deoxymyoglobin by bathing the gel in a diluting solution. Then the gel is introduced into a sample of dissolved oxygen in water. Dissolved oxygen reacts with deoxymyoglobin to give oxymyoglobin. The observed spectroscopic changes are indicative of dissolved oxygen in water. The highly porous coating can extract hydrophobic analytes (Lu *et al.* 1996) such as benzene from aqueous solutions and concentrate them inside the film. Periplasmic nitrate reductase (NAR) extracted from a denitrifying bacterium can be encapsulated in a sol–gel (Aylott *et al.* 1997). The enzyme reacts specifically with the nitrate (NO_3^-) anion. The reduction of nitrate by periplasmic nitrate reductase results in a characteristic change in UV–visible absorption of the nitrate reductase. Doped sol–gel crystals (Diaz and Peinado 1997) were derived from tetramethyl orthosilicate (TMOS) with cholinesterase using microencapsulation. Organophosphorus pesticides (ORPs) can be determined by fluorimetric detection based on inhibition of immobilized cholinesterase with the percentage of inhibition being correlated to pesticide concentration. It has been shown (Maitra *et al.* 1998) that a sol–gel material can be used for controlled drug delivery. The authors prepared injectable sprayable nanometer-size hydrated silica particles encapsulating a high-molecular-weight compound such as tyraminylinulin, FITC-dextran and horseradish peroxidase. The entrapment increased the efficiency of the entrapped molecules by about 80%. The entrapped compounds had zero leachability. Silicon-based biosensors have been developed (Ostraff *et al.* 1999) that generate a visual signal in response to nucleic acid targets.

A sol–gel derived porous silica carrier has been described (Santos *et al.* 1999) for the controlled release of proteins. The materials are intended to serve as both substrate for bone growth as well as to allow incorporated proteins such as growth factors to diffuse out and stimulate cell function and tissue healing. A sol–gel-based chemiluminescent H_2O_2 sensor has been reported. A new process was suggested (K.-M. Wang *et al.* 2000) to prepare a crack-free sol–gel membrane. The sensor showed a linear response towards H_2O_2 with a detection limit of 8 μM. An optical sensor that is highly sensitive to hydrogen peroxide (Lobnik and Cajlakovic 2001) has been developed. It was prepared by incorporating the indicator dye Meldola Blue (MB) into sol–gel layers that were prepared from tetramethyl orthosilicate (TMOS) and methyltrimethoxysilane (Me–Tri–MOS). Figure 3 schematically illustrates the formation of product in a sol–gel-based enzyme electrode with an entrapped mediator and with the mediator immobilized on the surface of an enzyme electrode.

5. SOL–GEL-DERIVED GLASSES FOR CLINICALLY IMPORTANT BIOSENSORS

The determination of the concentration of metabolites is important in clinical biochemistry. In most cases, the concerned metabolites in blood and urine are in the micro- and millimolar concentration range. A better understanding of several diseases requires accurate and quick measurements of steroids, drugs and metabolites present in blood serum. These substances are particularly important in intensive care medicine, surgery, and life-threatening situations. Since they are present in low concentration, it is not possible to accurately determine them with conventional methods. However they can be determined using immunoassays with the help of immobilized enzymes. Various methods have been employed over the last 10 years for the stabilization and immobilization of enzymes employed. Many research papers have been published relating to enzyme entrapment within a sol–gel matrix in an active form, which could be used for making biocatalysts or biosensors (Scheller and Schmid 1992). It is advantageous to use sol–gels as matrix for immobilization of enzymes due to their many favorable properties. Sol–gel-derived porous silica carriers for the controlled release of protein were synthesized using a room-temperature process (Santos *et al.* 1999). The advantages of immobilizing enzymes in sol–gels are that enzymes can be physically trapped without covalent bonding, which could lead to denaturation of active sites. The kinetics of sol–gel-immobilized enzymes closely resemble that of their water-soluble counterparts, and the size of the pores can be controlled in order to avoid leaching. Sol–gel matrices are stable against heat, chemicals and photo-degradation and their protective action enhances the stability of encapsulated enzymes. Sol–gel matrices can be obtained in a variety of forms (optically transparent glasses films, fibres, microspheres) and thin sol–gel films can be deposited onto most other materials like plastics, paper, metal, glasses, cloth, etc. The chemistry of protein in these gels is same as that in solution and the silica matrix inhibits the intermolecular interaction of encapsulated macromolecules. Finally, the presence of the rigid silica cage prevents protein movements. The advantage of this situation is that the reactivity inside the gel matrix occurs at the level of individual protein molecules.

A sensor based on the action of enzyme on glucose has been shown to be a better choice for the rapid determination of glucose. In this case, the conversion of glucose into gluconic acid by glucose oxidase has been utilized.

5.1. Glucose Biosensor

Glucose biosensors are used in the medical and scientific fields to achieve high selectivity and sensitivity for glucose molecules. The determination of glucose is the most frequently performed routine analysis for diabetics, as well as having uses in the food and fermentation industries.

The selective determination of blood glucose is very important for the screening and treatment of diabetes. The normal range of glucose in blood serum is in between 4.2 and 5.2 mmol/L and the normal concentration in urine is only 0.2 mmol/L (Scheller and Schmid 1992). The basic reaction involved is

$$\text{glucose} + O_2 \xrightarrow{\text{glucose oxidase}} \text{gluconic acid} + H_2O_2. \tag{3}$$

Researchers are looking for new materials that will permit more effective communication in sensor technology to meet the desired requirements of accurate and real-time information. The sol–gel technique is used to obtain a uniform pore size to allow the analyte species to enter the pore and interact with the immobilized enzyme. The coating is preferably made of various materials such as silica, zirconia, or alumina. For the fabrication of glucose biosensors mostly researchers have used tetramethyl orthosilicate (TMOS). The pH of the material is close to neutral. The determination of blood sugar is a routine test in clinical diagnostics.

Glucose biosensors have been used for analysis in clinical laboratories, food industries and in microbiology. A variety of methods are available for the fabrication of enzyme electrodes.

Several reports are available on the immobilization of enzymes, and glucose oxidase in particular (Kraus *et al.* 1993). A glucose biosensor consisting of a tetramethyl orthosilicate sol–gel system containing immobilised glucose oxidase was reported (Tatsu *et al.* 1992). This biosensor utilized an $8 \times 2\,\text{mm}^2$ disk of tetramethylorthosilicate-(TMOS-) derived xerogel doped with glucose oxidase, peroxidase and a chromogenic dye for the detection of glucose. A tetraethylorthosilicate-based sol–gel monolith doped with glucose oxidase was prepared by Tatsu *et al.* (1992). The performance of the glucose sensing element monolith using flow injection analysis demonstrated a response time of about 4 minutes. The activity of encapsulated glucose oxidase was investigated by a photometric method. Platinum electrodes coated with a sol–gel-derived film of V_2O_5 doped with GOD have been used as glucose biosensors for the estimation of glucose concentration (Lev and Glezer 1993) using cyclic voltammetry. The composite benefit arose both from the porosity and the rigidity of the silica matrix and the electrical

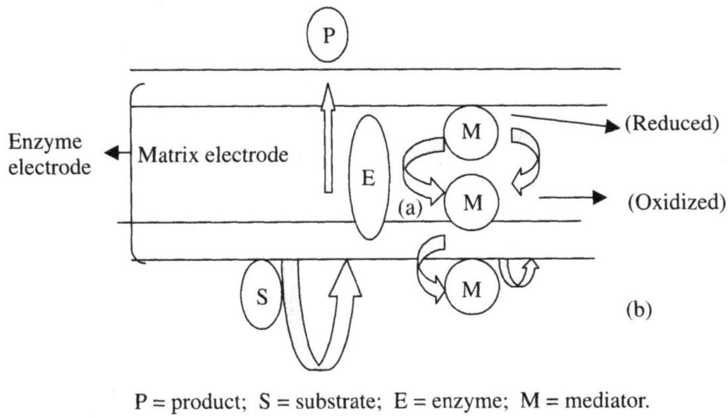

Figure 2. Schematic of the formation of product in an enzyme electrode, (a) when the mediator is entrapped within the matrix along with the enzyme and (b) when the mediator is immobilized on the surface of an enzyme electrode.

conductivity of the graphite. Glucose oxidase (GOD) was first adsorbed on the surface of the carbon powder and then used for the preparation of the composite sol–gel film on a glassy-carbon electrode.

A ferrocene-mediated sol–gel biosensor was constructed by Audebert *et al.* (1993) using a two-stage sol–gel preparation procedure based on tetramethyl orthosilicate (TMOS) and commercial colloidal silica of varying particle size (Figure 2). These authors have shown that more than 80% of GOD retains its activity in the gel and the Faradic response of the electrode agrees well with theoretical calculations based on this activity. Sol–gel derived films with ultrafine pore size and very high surface area have bee·, produced by Narang *et al.* (1994b). Their unique nanostructure is responsible for the unusual optical, thermal, electrical and many other properties. These authors described a tetraethylorthosilicate-derived thin film fabricated by a sol–gel process and doped with glucose oxidase used for the estimation of glucose concentration. Mediator- (ferrocene-) based sol–gel biosensors have also been developed for the estimation of glucose concentration (Sampath and Lev 1996) (Figure 2). The activity of glucose oxidase immobilized within SiO_2 gels is mainly determined by the chemical and physical parameters of the sol–gel process (Kunzelmann and Bottcher 1997). Both the optical and electrochemical methods (Figure 3) have been used to measure the activity of immobilized glucose oxidase. The modified glucose oxidase

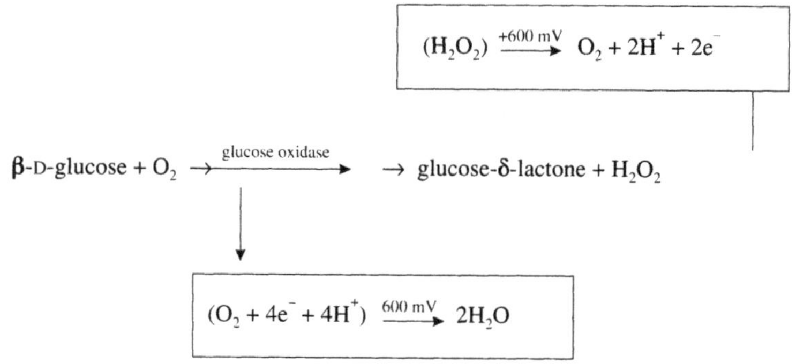

Figure 3. Schematic of regeneration of oxygen (O_2) during the estimation of glucose with GOD sensor.

facilitates communication between the enzyme and the conducting graphite in the sol–gel matrix. Sol–gel-derived inks containing surfactant bis(2-ethylhexyl)sulfosucceinate have been reported. These have been shown to be useful for eliminating the need for co-solvent and acid catalyst. A glucose microsensor has been described based on the coverage of photolithographically microfabricated silicon wafer with glucose-oxidase-doped tetraethylorthosilicate sol–gel film prepared in two steps. A Nafion overlayer was used to avoid the effects of interferent. A non-mediated glucose biosensor based on encapsulated glucose oxidase (GOD) within the composite sol–gel glass was fabricated by Pandey *et al.* (1999a) using an optimum concentration of 3-aminopropyltriethyl silane 2-(3,4-epoxycyclohexyl)ethyltrimethoxysilane, glucose oxidase dissolved in double distilled water and HCl. Fast responding and sensitive glucose biosensors have been fabricated recently using a sandwich configuration with sol–gel organic–inorganic hybrid materials (Pandey *et al.* 1999b).

Bharathi and Lev (1998) introduced sol–gel-derived nanocrystalline gold–silicate composite electrochemical biosensors and demonstrated their use for glucose sensing. A new type of inorganic biosensor was introduced (Bharathi and Lev 2000). The sensor comsisted of glucose oxidase encapsulated in a sol–gel-derived Prussian blue–silicate hybrid network. Glucose was detected by the biocatalytic reduction of oxygen followed by catalytic reduction of hydrogen peroxide by the Prussian blue catalyst. The sol–gel silicate entails a rigid encapsulating matrix, the Prussian blue provides chemical catalysis and charge mediation from the reduction site to the supporting electrode, and the enzyme is responsible for the biocatalysis. They also demonstrated the feasibility of a dual optical/electrochemical mode of analysis.

A microdialysis sensor system for continuous glucose measurements has been developed by Yu and Chen (2000). The sensor is based on co-immobilizing glucose oxidase (GOD) with the peroxidase on the surface of the silicon base with an Au interdigited electrode using the tetramethyl orthosilicate (TMOS) sol–gel technique. A layered ("sandwich") immobilizing method is used to improve the stability and the lifetime of the co-enzyme system. The result of the optimized co-immobilized sensor system demonstrated good agreement for between the glucose concentration over the linear range 0.5–50 mM, and the response time is of the order of less than 2 min. Wolfbeis *et al.* (2000) developed various types of thin-film glucose biosensors using oxygen transducer and sol–gel immobilised glucose oxidase.

Recently, Kros *et al.* (2001) have described the preparation of sol–gel silica-based biocompatible coatings which can be used for future implantable glucose sensors. TEOS was used as precursor for water-borne silicate gels of which the properties were varied by mixing the sol with polyethylene glycol (SG-PEG), heparin (SG-HEP), dextran sulfate (SG-DS), Nafion (SG-NAF) or polystyrene sulfonate (SG-PSS). Glucose measurements using glucose-oxidase-based sensors coated with different hybrid films were performed both in buffered solutions containing bovine serum albumin and in serum. They obtained stable glucose responses for the coated sensors in both media. The coating containing SG-DS appeared to be the most promising for future *in vivo* glucose measurements.

A sol–gel-based glucose biosensor using thermometric measurement was reported by Ramanathan *et al.* (2001). The enzymes (glucose oxidase, GOD, and catalase, CAT) were entrapped on the surface of reticulated vitreous carbon cylinder (RVC cartridge) using SG as a binder. This 'RVC cartridge' was placed within the column of an enzyme thermistor (ET) device. Injection of various D-glucose concentrations resulted in different heat content of the circulating buffer, recorded as a thermometric peak by a sensitive thermistor. The stability of the entrapped GOD/CAT stored at room temperature (25°C) or 4–10°C was 3 or 6 months, respectively. The effect of dissolved oxygen and other interferents, such as acetaminophen, ascorbic acid, aspartic acid, glutamic acid, urea and uric acid, on the catalytic activity of the enzyme was also investigated. This system was employed to detect glucose in samples of fruit juice, Cola drinks and human blood serum.

Ulatowska *et al.* (2001) used sol–gel technology to produce glass-like materials doped with different compounds. The sol–gel matrices were prepared using acid hydrolysis. They prepared materials in the form of thin layers. The structure was visualized by means of a light microscope. The microscopic images were digitized and their luminances were calculated. The thin-film-type matrices were investigated also by means of small-angle light scattering. The material properties with dopants such as albumin, ruthenium complex and osmium tetraoxide were examined. Some matrices showed more homogeneous structure. Lower scattering in doped materials than in the pure matrices was observed.

Liu *et al.* (1999) reported a sol–gel-derived amperometric biosensor for glucose. Glucose oxidase was immobilized in the Al_2O_3 sol–gel matrix on a platinized glassy carbon electrode. A novel amperometric glucose biosensor was constructed by electrochemical formation of a polypyrrole (PPy) membrane in the presence of glucose oxidase (GOD) on the surface of a horseradish-peroxidase- (HRP-) modified ferrocenecarboxylic-acid- (FCA-) mediated sol–gel-derived ceramic carbon electrode. The amperometric detection of glucose was carried out at +0.16 V (vs. SCE) in 0.1 mol/L phosphate buffer solution (pH 6.9) with a linear response range between 8.0×10^{-5} and 1.3×10^{-3} mol/L of glucose. The biosensor showed a good suppression of interference and a negligible deviation in the amperometric detection.

5.2. Urea Biosensor

The concentration of urea in blood is an important parameter in clinical chemistry. Ammonium ions and urea play important role in analytical and clinical chemistry being linked to many processes such as blood and urine analysis, kidney failure and artificial kidney control. Ammonium ion concentration increases in athletes during endurance training and performance. Ammonium ion and urea determination are also important in food and drug analysis. The normal range of urea concentration in blood serum is 3.6–8.9 mmol/L (Ho *et al.* 1999). Urease catalyzes the hydrolysis of urea to produce ammonia, which in turn reacts with Nessler's reagent to form a colored product. By following the absorbance of the coloured product at 405 nm, urea can be quantified. The key reaction is

$$NH_2\text{–}CO\text{–}NH_2 + H_2O \xrightarrow{\text{urease}} 2NH_4^+ + OH^- + HCO_3^-. \tag{4}$$

Various urease-based urea sensors have been reported (Scheller and Schmid 1992). A novel protocol to entrap active urease in tetraethyl orthosilicate-derived sol–gel film has been described by Narang *et al.* (1994a). A urease-fixed sol–gel silica electrode has been applied to the potentiometric determination of urea (Ogura *et al.* 1999). Urease has been immobilized in a sol–gel silica matrix, and the enzyme-fixed silica characterized by means of thermogravimetry/mass spectrometry and Fourier transform infrared (FTIR) spectroscopy. The former indicated that the urease immobilized in the sol–gel silica is thermally more stable than free urease. The FTIR spectrum of the urease-fixed silica results from the superimposition of the silica and urease spectra, confirming that the enzyme can be immobilized in the sol–gel silica without any chemical deterioration. A Nernstian relationship (slope 59 mV/decade) was obtained for urea concentration lying between 1×10^{-5} and 5×10^{-2} M at pH 8.

A conductometric biosensor for urea determination in serum based on sol–gel-immobilized urease on a screen-printed interdigited array (IDA) electrode has been developed by Lee *et al.* (2000). They reported that the screen-printed thick-film IDA electrode is an excellent conductometric transducer in which the admittance signal is dominated by the conductance signal and the resulting dynamic range is wide enough to be employed for the construction of a conductometric urea biosensor. Urease was immobilized by a conventional sol–gel process using tetramethyl orthosilicate as a precursor. The sol–gel-derived urea biosensor showed a reasonably wide linear dynamic range of 0.03–2.5 mM in 5.0 mM imidazole–HCl buffer at pH 7.5 (detection limit 30 μM, signal/noise ratio 2), thus enabling it to be used to monitor urea in a diluted serum sample with the differential measurement format consisting of an active IDA electrode with sol–gel-immobilized urease and a control IDA electrode with sol–gel-immobilized bovine serum albumin. The urea biosensor exhibited good sensor-to-sensor reproducibility (4.4%) and storage stability (63% of its original activity retained after 25 days). The coupling of the sol–gel immobilization method and screen-printing technology can offer mass production of a durable urea biosensor.

Urea sensors based on the pH dependence of the anodic oxidation of hydrazine have also been utilized in the glucometer for haemodialysis monitoring. An efficient urea analyzer has been developed for undiluted blood samples by using a urease-covered ammonium-ion selective electrode in an FIA system. A new solid-state pH sensor has been developed using neutral poly(3-cyclohexylthiophene) assembled over a Pt disk electrode (Pandey *et al.* 2000). The polymer-modified electrode was sensitive to pH and a reversible super-Nernstian behavior was observed. The polymer-coated electrode was subsequently used to construct an all-solid-state urea sensor. The construction of this new urea sensor involved two major steps; (a) 20 uL of urease solution (40 mg/mL) was allowed to assemble overnight at 4°C over a neutral poly(3-cyclohexylthiophene)-modified electrode; (b) an organically modified sol–gel layer was allowed to form over the urease adsorbed polymer modified electrode. The new solid-state urea sensor provided excellent measurement reproducibility and was stable for 3 months when stored at 4°C under dry conditions.

A reactor containing urease bound to porous glass in a flow injection analysis system was reported by Mosbach (1988). The product was detected by means of an ammonia electrode after alkalization of the measuring solution. An integrated optical sensor bearing a miniature flow-through cell with a urease reactor has also been fabricated. To construct such a sensor a planar light-conducting SiO_2–TiO_2 structure was patterned by photolithography and etched in a grid relief on a Pyrex support. The sensor functioned as a differential refractometer indicating changes of index of refraction at the sensor surface as well as the thickness of layers absorbed on the grid region. Its applicability to urea measurement has been demonstrated. A sol–gel-based optochemical sensor for ammonia has also been described (Trinkel *et al.* 1996). The sol–gel entrapment played an important role in stabilizing urease. Preparation of a urea electrode using perfluoroalkylated enzyme, urease, immobilized for use with an ammonia gas-sensing electrode has been reported. A new sol–gel/enzyme/sol–gel sandwich architecture based on the enzyme urease (hexamer, Mw = 590 000) encapsulated between two sol–gel-derived thin films has also been reported (Kumar *et al.* 1994). This new film architecture was used as a sensing element for the quantification of urea. The activity, detection limits, linear dynamic range, preparation repeatability and performance, storage and operation stability, and response time of this new sensor platform were characterized. The response of the new sensing element was followed by using the well-known complexation of Nessler's reagent. Ormosils were prepared (Lobnik and Wolfbeis 1998) in different ratios of tetramethyl orthosilicate and organically modified sol–gel precursors of type $RSi(OR)_3$ and $R_2Si(OR)_2$. The pH indicator containing aminofluorescein was incorporated into various ormosils and the resulting materials were tested for their response to pH and dissolved ammonia at constant pH. This could be an enzyme-less approach for the estimation of urea.

5.3. Cholesterol Biosensor

Cholesterol, the most abundant steroid in humans, is a component of the cell membranes of most body tissues and is widely distributed throughout the human body, including the cerebral nerves and adrenal glands. It plays an important role as a hormone precursor. Some cholesterol is ingested with food and the rest is synthesized in the liver. Cholesterol, which exists in blood mostly in the esterified form, is combined with specific proteins and is carried with other lipids. 'Blood total cholesterol' in clinical tests indicates the total concentration of non-esterified cholesterol plus esterified cholesterol in the blood. It occurs in free form at about 30% as well as in esterified form with fatty acid at about 70% (Scheller and Schmid 1992). The assay of cholesterol is important for the diagnosis of disorders in lipid metabolism. Cholesterol determination in blood is known to be clinically important for diagnosis of heart diseases, arteriosclerosis and cerebral thrombosis. The normal concentration of cholesterol in blood serum is 3.1–6.7 mmol/L (Ho *et al.* 1999). The concentration of cholesterol is also important in fermentation and pharmaceutical industry.

The basis of all enzymatic total cholesterol assays is the hydrolysis of cholesterol ester by cholesterol esterase to free cholesterol and fatty acid. Cholesterol is further oxidized with the help of cholesterol oxidase:

$$\text{cholesterol ester} \xrightarrow{\text{cholesterol esterase}} \text{cholesterol} + \text{fatty acids}, \tag{5}$$

$$\text{cholesterol} + O_2 + \text{ChOx} \rightarrow \text{cholest-4-en-3-one} + H_2O_2. \tag{6}$$

The H_2O_2 can be detected amperometrically as well as by using a dye such as 4-aminoantipyrine. The reaction proceeds as follows:

$$\begin{aligned} &H_2O_2 + \text{phenol} + \text{4-aminoantipyrine HRP} \\ &\quad \rightarrow (p\text{-benzoquinone monoimino})\text{phenazone} + 4H_2O. \end{aligned} \tag{7}$$

For analytical purpose the cholesterol oxidase has been immobilized on various carriers. Cholesterol oxidase has been immobilized on a collagen membrane, and was also immobilized on a nylon membrane (Mosbach 1988) by covalent linkage and fixed onto an O_2 probe. Such a sensor has been used for the estimation of cholesterol concentration. A cholesterol sensor has also been designed by absorbing NMP^+ $TCNQ^-$ HRP on an organic metal electrode and covering it by nylon-bound cholesterol oxidase. The use of mono- or bi-alkylated alkoxysilane makes the gels more hydrophobic and suitable for lipophilic analysis (Reetz *et al.* 1995 and Kuncova *et al.* 1995).

A sol–gel silicate-based biosensor for cholesterol, glucose and galactose was made by utilizing a composite membrane of sol–gel enzyme film and electrochemically generated poly(1,2-diaminobenzene) film to improve the selectivity of the sol–gel enzyme sensors (Yao and Takashima 1998). The stability of the sensor was improved by exposing the enzyme layer to glutaraldehyde vapor. Cholesterol oxidase (ChOx)/ horseradish peroxidase (HRP) was immobilized in thin sol–gel films derived from tetraethylorthosilicate (TEOS) (Kumar *et al.* 2000) by physical adsorption, microencapsulation and physical entrapment [sol–gel/enzyme/sol–gel] sandwich techniques. The response studies of these ChOx/HRP-immobilized TEOS sol–gel films were carried out by both spectrophotometric and amperometric methods. The results obtained from spectrophotometric and electrochemical measurements using these systems yield linearity over the range 2–10 mM cholesterol. The UV–visible absorption at 240 nm of enzymatically formed cholestenone was used as the measuring signal. All enzyme sol–gel films are found to be stable for about 8 weeks at 25°C and 12 weeks at 4–5°C. Sol–gel film coated with bovine serum albumin was used to covalently link cholesterol oxidase to enhance the sensitivity of sensors towards cholesterol (Kumar *et al.* 2001). The BSA–ITO (bovine serum albumin–Chox–indium tin oxide coated glass plate) sol–gel film was characterized by spectroscopic and SEM techniques. The response time of this cholesterol biosensing electrode was found to be 60 seconds and it was stable for about 12 weeks.

5.4. Lactate Biosensor

Lactate determination is important in the diagnosis of respiratory insufficiencies and myocardial disorders and in neonatology and sports

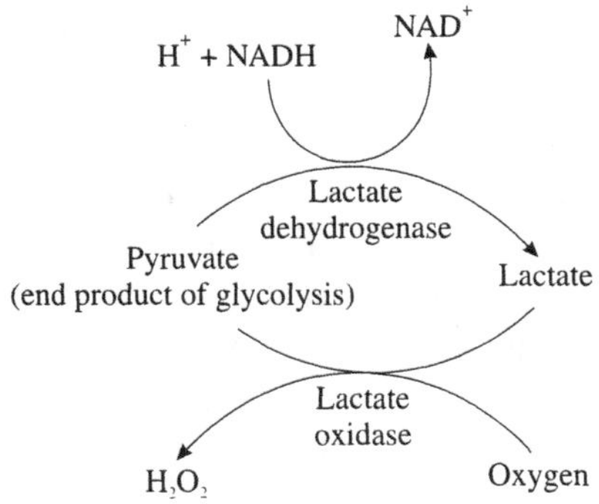

Figure 4. Reaction scheme for lactate oxidase and lactate dehydrogenase.

medicine. Its determination is also useful in food, dairy industries and sports medicine. Besides this, biosensors have been considered important in healthcare, veterinary medicine, fermentation process etc. Among all these, technical development of biosensors for medical care has demanded the greatest attention. The major advantage of this technology lies in its exploitation of biological specificity, which facilitates the ability of the device to discriminate low concentrations of analyte in complex matrices. The normal lactate concentration in blood lies between 1.2 and 2.7 mmol/L. Strenuous exercise may raise lactate levels to 25 mmol/L (Scheller and Schmid 1992). For accurate lactate determination, heamolysis of the sample is required, and the glycolytic reactions in the sample have to be efficiently and rapidly inhibited in order to avoid lactate formation. The best sample for lactate determination is deproteinized blood. Efforts are being made to develop lactate sensors by different methods that may be readily used at the bedside. One widely used enzymatic reaction for lactate determination is catalyzed by lactate dehydrogenase:

$$\text{lactate} + \text{NAD} \xrightarrow{\text{LDH}} \text{NADH} + \text{pyruvate}. \tag{8}$$

[NAD, nicotinamide dinucleotide; LDH, lactate dehydrogenase; NADH, reduced NAD.]

The reaction scheme is given in Figure 4.

Stability and longevity of chemical and biological sensors is a major issue restricting wider application of the technology. NMRC (National Microelectronic Research Centre, Ireland) has been researching methods to improve the lifetime of biosensors for several years and has developed a sol–gel-based lactate biosensor with excellent longevity (NMRC report,

1999). Lactate oxidase is known to be a difficult enzyme to stabilise. The NMRC results show that an optimised formulation of the sol–gel matrix can yield lactate oxidase biosensors with lifetimes exceeding 3 months and with sensitivities and response times as good or better than alternative technology devices reported in the literature.

An improved blood sample pretreatment was carried out by addition of cetyltrimethyl ammonium bromide. This compound haemolyses the blood sample and stabilizes the lactate content. Lactate oxidase has been covalently bound by 3-aminopropyltriethoxysilane and cross-linked by glutaraldehyde (Anzai *et al.* 1992). This membrane was found to be highly selective for hydrogen peroxide. Attempts have been made to measure lactate/pyruvate in plasma by using a lactate dehydrogenase–LMO sequence electrode (Tsuchida *et al.* 1985). The sensor was connected to a pO_2 meter and was equally sensitive for lactate and pyruvate. This group described an amperometric biosensor for lactate constructed by first casting lactate oxidase and an osmium redox mediator onto the glassy carbon substrate, followed by coverage with sol–gel film derived from methyltriethoxysilane (MTEOS). The advantage was the inorganic character of the sol–gel immobilization matrix for the biosensing of peroxide in non-aqueous media (Livage 1996). The immobilization of LDH on tetraethylorthosilicate- (TEOS-) derived sol–gel films (Ramanathan *et al.* 1997) including other studies on the enzyme assays as a function of film thickness, time, pH and pyruvate concentration have been reported. Lactate dehydrogenase (LDH) (Chaubey *et al.* 2001) was immobilized on TEOS-derived sol–gel films on glass plates by physical adsorption and the sandwich configuration. Once again the merits and demerits of these methods of immobilization under varying concentration of process parameters like pH, temperature, different buffer systems, lactate concentration etc. have been investigated.

5.5. Miscellaneous Applications

Apart from the above-mentioned uses, sol–gel glasses find several other important applications. Recently Lam *et al.* (2001) have developed a phosphorescence-based oxygen sensor based on silica glass doped with erythrosin B.

Ceramics and glasses have been widely investigated as materials for the design of artificial bone graft substitutes (Damien and Parsons 1991).

Santos *et al.* (1999) synthesised sol–gel-derived porous silica carriers for controlled release of protein both to serve as substrates for bone growth and to allow incorporated proteins such as growth factors to diffuse out and stimulate cell function and tissue healing. The characteristics of the protein release exhibited by these sol–gel materials represent the basis to pursue the application of the delivery of biologically active molecules under clinically relevant conditions (Jain *et al.* 1998, Thust *et al.* 1996).

The measurement of blood pH is one of a number of clinical markers that have been identified to assess the health of blood-vessel tissue at the site of a stroke, thereby providing guidance in stroke therapy. When tissue dies, lactic acid builds up and blood pH decreases. So if blood pH is below normal (7.4) at or near the stroke site, then brain cell death has occurred, and the use of neuroprotectant drugs to minimise brain damage is unwarranted. If on the other hand pH is close to normal, cell death has not occurred, and neuroprotectant drugs become a therapeutic option (Zauner *et al.* 1995, Patrick 1997). A local blood pH sensor based on the sol–gel technique has been reported by Grant and Glass (1997). Recently, Grant *et al.* (2001) have reported a fibre-optic sensor that can be used to measure pH in brain tissue. The fibre-optic sensor can be threaded through a catheter up the vascular system to the affected region of the brain of stroke patients or patients with traumatic head injury. They utilized the sol–gel technique to entrap the pH-sensitive dye, semi-naphthorhodamine-1 carboxylate (SNARF-1C), at the tip of an optic fibre. These sensors were tested *in vitro*. *In-vitro* results showed linear and reproducible responses in human blood in the pH range 6.8–8.0.

Pandey *et al.* (2001) reported an electrocatalytic ethanol biosensor using ferrocene-encapsulated palladium- (Pd-) linked organically modified sol–gel glass (ormosil). The electrochemistry of this ormosil was found to be reversible and is promising for the development of an ethanol/alcohol biosensor. This biosensor was found to have fast response time and high reproducibility.

Flora and Brennan (1998) reported the development of a fluorometric detection strategy for Ca^{2+} based on induced changes in the conformation of cod III parvalbumin entrapped within a sol–gel-processed glass. The detection scheme utilizes a fluorescent allosteric signal transduction (FAST) strategy wherein conformational changes induced by Ca^{2+} binding result in alterations in the intrinsic fluorescence from the single tryptophan residue. Entrapment caused improvements in protein stability against chemical denaturants. Moreover, there was no interference from divalent ions such as Mg^{2+}, Sr^{2+} or Cd^{2+}, indicating the viability of using sol–gel-entrapped FAST proteins for the detection of Ca^{2+}.

Kane *et al.* (1998) constructed sol–gel-coated horseradish peroxidase (HRP) electrodes by complexing HRP on a glassy carbon electrode with the redox osmium polymer [Os(bpy)(2)(PVP)(10)Cl]Cl. Results were reported for the determination of phenols in both aqueous and organic phase systems. The biosensor exhibited a low operating potential, i.e. 0 V versus Ag/AgCl, fast response times, high sensitivity and micromolar detection limits for a range of selected phenols.

Amperometric enzyme electrodes for phenolic compounds via an easy and effective immobilization method using the sol–gel silica have been reported (Li *et al.* 1998, B. Wang *et al.* 2000). The enzyme electrode comprises tyrosinase immobilized in thin sol–gel silica. The tyrosinase retains its bioactivity when being immobilized by the sol–gel film.

An amperometric sol–gel-based enzyme inhibition electrode for the detection of low levels of cyanide was constructed by immobilizing horseradish peroxidase (HRP) and an osmium redox polymer ([Os(bpy)(2)(PVP)(10)Cl]Cl, Os-polymer) as mediator (Park *et al.* 1997). Upon addition of hydrogen peroxide to the solution, a bioelectrocatalytic reduction wave was observed, which was diffusion controlled. This reduction current was subsequently inhibited by cyanide. The steady-state hydrogen peroxide catalytic reduction current reached a plateau at 150 mV (vs. Ag/AgCl) and cyanide could be determined amperometrically with good sensitivity at 0 mV (vs. Ag/AgCl).

Specific antigen–antibody reactions have been performed within sol–gel matrices, extending the field of sol–gel chemistry toward immunosensors (Santos *et al.* 1999). Antigens are macromolecular substances recognized by the immune system as foreign. Small molecules have to be bound to a macromolecular carrier to induce an immune response and stimulate the production of antibodies. These antibodies have two fixation sites that can bind antigens with a high level of specificity. Antibodies have been encapsulated within sol–gel matrices by R. Wang *et al.* (1993).

6. CONCLUSIONS

Sol–gel-based sensing materials developed to date have served primarily as microsphere support matrices for recognition molecules. It has been shown that sol–gel chemistry can provide an alternative versatile and simple route for the development of electrochemical and optical biosensors for many types of clinical analysis. The biological application of sol–gel chemistry is quite a new field. The doped glass developed by the sol–gel process can be used as an enzyme electrode for the estimation of different concentrations of glucose, cholesterol, urea, lactate and various other metabolites in solution.

While most of the sol–gel-based biosensors reported to date have relied on the encapsulation of enzymes, new biosensors based on immobilization of antibodies, whole cell, nucleic acid, receptors are anticipated in the near future. Sensor arrays containing multiple microelectrodes and biogels are expected, in view of the compatibility of sol–gel method with different biological elements. Recent research in the area is very encouraging but a considerable amount of fundamental research remains to be done. A better understanding of the correlation between the biomolecule, the reaction conditions and the properties of sol–gel-based material should lead to a great potential for the applications of sol–gel glasses in biosensors. In addition, the chemistry of silicon and non-silicon alkoxide is likely to add new degrees of utility to sol–gel-based sensing electrodes. There is no doubt that doped glasses developed by sol–gel processes containing biomolecules will play a major role in future biosensing technology.

ACKNOWLEDGEMENTS

We are thankful to Dr. K. Lal, Director, National Physical Laboratory, New Delhi, India, for the constant encouragement and his interest in this work. We thank Dr. S.S. Bawa, Dr. K.K. Saini, Ravi Ranjan Pandey, M.K. Pandey and Ms Asha Chaubey for interesting discussions. The financial support received under the Department of Biotechnology project (DBT) (DP/CHBIO/CLP/003832), Govt of India, New Delhi India is gratefully acknowledged.

REFERENCES

Anzai, J., N. Inomata, T. Osa, M. Morita and K. Maehare (1992). A facile method of regulating the dynamic range of lactate sensor based on polymer coating. *Chem. Pharm. Bull.*, **40**, 2897–2899.

Audebert, P., C. Decmaille and C. Sanchez (1993). Electrochemical probing of the activity of glucose oxidase embedded sol–gel matrices. *Chem. Mater.*, **5**, 911.

Avellaneda, C.O., M.A. Macedo, A.O. Florentino and M.A. Aegerter (1994). Sol–gel coatings for opto electronic devices. *Proc. SPIE*, **2255**, 38–51.

Avnir, D. (1995). Organic chemistry within ceramic matrices: doped sol–gel materials. *Acc. Chem. Res.*, **28**, 328–334.

Aylott, J.W., D.J. Richerdson and D.A. Russell (1997). Optical biosensing of nitrate ion using a sol–gel immobilized nitrate reductase. *Analyst*, **122**(January), 77–80.

Bharathi, S., and O. Lev (1998). Sol–gel derived nanocrystalline gold silicate composite biosensor. *Anal. Commun.*, **35**, 29–31.

Bharathi, S., and O. Lev (2000). Sol–gel-derived prussian blue-silicate amperometric glucose biosensor. *Appl. Biochem. Biotechnol.*, **89**(2–3), 209–216.

Braun, S., S. Shtelzer, S. Rappoport, R. Zusmann, D. Avnir and M. Ottoleaghi (1992). Biocatalyst by sol–gel entrapped enzymes. *J. Non-Cryst. Solids*, **147**, 739–743.

Brinker, C.J., and G.W. Scheterer (1990). *Sol–Gel Science*. Academic Press, New York.

Cauqui, M.A., and J.M. Rodriguez-Izquierdo (1992). Application of the sol–gel catalyst preparation. *J. Non-cryst. Solids*, **147–148**, 724–738.

Chaubey, A., M. Gerad, V.S. Singh and B.D. Malhotra (2001). Immobilization of lactate dehydrogenase on tetraethylorthosilicate-derived sol–gel film for application to lactate biosensor. Presented at Internatioanl Conference on 'New Horizons in Biotechnology', 18–21 April 2001, Trivandrum, India.

Damien, C.J., and J.R. Parsons (1991). Bone graft and bone graft substitutes: a review of current technology and applications. *J. Appl. Biomater.*, **2**, 187–208.

Dave, B.C., B. Dunn, J.S. Valentine and J.I. Zink (1994). Sol–gel encapsulation methods for biosensors. *Anal. Chem.*, **66**(22, November 15), 1120A–1127A.

Diaz, N.V., and M.C.R. Peinado (1997). Sol–gel cholinesterase biosensor for organophosphorus pesticide fluorimetric analysis. *Sensors Actuators B*, **38–39**, 426–431.

Ellerby, L.M., C.R. Nishida, F. Nishida, S.A. Yamanaka, B. Dunn, J.S. Valentine and J.I. Zink (1992). Encapsulation of protein in transparent porous silicate glasses prepared by sol–gel method. *Science*, **255**, 1113–1115.

Fegley Jr, B., and E.A. Barringer (1984). In *Better Ceramics through Chemistry*, Vol. 32. C.J. Brinker, D.E. Clark and D.R. Ulrich (Eds.), Elsevier, New York, p. 187.

Flora, K., and J.D. Brennan (1998). Fluorometric detection of Ca^{2+} based on an induced change in the conformation of sol–gel entrapped parvalbumin. *Anal. Chem.*, **70**(21), 4505–4513.

Grant, S.A., and R.S. Glass (1997). A sol–gel based fiber optic sensor for local blood pH measurements. *Sensors Actuators B*, **45**(1), 35.

Grant, S.A., K. Bettencourt, H.J. Krulevitch and R. Glass (2001). In vitro and in vivo measurements of fiber optic and electrochemical sensors to monitor brain tissue pH. *Sensors Actuators B*, **72**, 174–179.

Guilbault, G.G. (1984). *Analytical Uses of Immobilized Enzymes*. Marcel Dekker, New York, pp. 77–111.

Hardy, A., G. Gowda, T.J. McMohan, R.E. Riman, W.E. Rhina and H.K. Bowen (1988). In *Ultra-structure Processing of Advanced Ceramics*. J.D. Mackenzie and D.R. Ulrich (Eds.), Wiley, New York, p. 40.

Hench, L.L., and D.R. Ulrich (Eds.) (1984). *Ultrastructure Processing of Ceramics Glasses and Composites*. Wiley, New York.

Hench, L.L., and K. West (1990). The sol–gel process. *Chem. Rev.*, **90**, 33–72.

Ho, W.O., S. Krause, C.J. McNeil, J.A. Pritchard, R.D. Armstrong, D. Athey and K. Rawson (1999). Electrochemical sensor for measurement of urea and creatinine in serum based on AC impedance measurement of enzyme catalyzed polymer breakdown. *Anal. Chem.*, **71**, 1940–1946.

Jain, T.K., I.R. De and A. Maitra (1998). Nanometer silica particles encapsulating active compounds: a novel ceramic drug carrier. *J. Am. Chem Soc.*, **120**(43), 11092–11095.

Kane, S.A., E.I. Iwuoba and M.R. Smyth (1998). development of a sol–gel based amperometric biosensor for the determination of phenolics. *Analyst*, **123**(10), 2001–2006.

Klein, L. (1988). *Sol–gel Technology*. Noyes Publications, Park Ridge, IL.

Kraus, S.C., R. Czolk, J. Reichert and H.J. Ache (1993). Optimization of the sol–gel process for the development of optochemical sensor. *Sensors Actuators B*, **15–16**, 199–202.

Kros, A., M. Gerritsen, V.S.I. Sprakel, N.A.J.M. Sommerdijk, J.A. Jansen and R.J.M. Nolte (2001). silica-based hybrid materials as biocompatible coatings for glucose sensors. *Sensors Actuators B*, **81**(1), 68–75.

Kumar, A., N.D. Kumar, M.N. Kamalasanan, B.D. Malhotra, S. Chandra, U. Narang, P.N. Prasad and F.V. Bright (1994). A novel protocol to entrap active urease in a tetraethoxysilane-derived sol–gel film by. Chem. Mater, **6**, 1596–15998.

Kumar, A., Rajesh, B.D. Malhotra and S.K. Grover (2000). Co-immobilization of cholesterol oxidase peroxidase in sol–gel film. *Anal. Chim. Acta*, **414**, 43–50.

Kumar, A., R.R. Pandey, K.K. Saini and B.D. Malhotra (2001). Covalent coupling of cholesterol oxidase to bovine serum albumin coated sol–gel film, In *Conference on Biotechnology – The Science & the Business*, September 28–30. IIT New Delhi, India, p. 73.

Kuncova, G., M. Guglielmi, P. Dubina and B. Safar (1995). Lipase immobilized by sol–gel technique in layers. *Ccollect. Czech Chem. Commun.*, **60**, 1573–1577.

Kunzelmann, U., and H. Bottcher (1997). Biosensor properties of glucose oxidase immobilized within SiO_2 gels. *Sensors Actuators B*, **38–39**, 222–228.

Lam, S.K., M.A. Chan and D. Lo (2001). Characterisation of phosphorescence oxygen sensor based on erythrosin B in sol–gel silica in wide pressure and temperature ranges. *Sensors Actuators B*, **81**, 135–141.

Lee, W.Y., S.R. Kim, T.H. Kim, K.S. Lee, M.C. Shin and J.K. Park (2000). Sol–gel-derived thick-film conductometric biosensor for urea determination in serum. *Anal. Chim. Acta*, **404**, 195–203.

Lev, O., and V. Glezer (1993). Sol–gel vanadium pentaoxide glucose biosensor. *J. Am. Chem. Soc.*, **115**, 2533.

Li, J., L.S. Chia, N.K. Goh and S.N. Tan (1998). Silica sol–gel immobilized amperometric biosensor for the determination of phenolic compounds. *Anal. Chim. Acta*, **362**(2–3), 203–211.

Liu, Z., B. Liu, M. Zhang, J. Kong and J. Deng (1999). Al_2O_3 sol–gel derived amperometric biosesor for glucose. *Anal. Chim. Acta*, **92**, 135–141.

Livage, J. (1996). Bioactivity in sol–gel glasses. *C.R. Acad. Sci. Paris Ser. IIb*, pp. 417–427.

Lobnik, A., and M. Cajlakovic (2001). Sol–gel based optical sensor for continuous determination of dissolved hydrogen peroxide. *Sensors Actuators B*, **74**, 194–199.

Lobnik, A., and O.S. Wolfbeis (1998). Sol–gel based optical sensor for dissolved ammonia. *Sensors Actuators B*, **51**, 203–207.

Lu, Y., L. Han, C.J. Brinker, T.M. Niemczyk and G.P. Lopez (1996). Chemical sensors based on hydrophobic porous sol–gel films and ATR-FTIR spectroscopy. *Sensors Actuators B*, **35–36**, 517–521.

Maitra, A.N., et al. (1998). See http://www.serc-dst.org/Chemicalprojects.htm.

Matijevic, E. (1988). In *Ultra-Structure Processing of Advanced Ceramics*. J.D. Mackenzie and D.R. Ulrich (Eds.), Wiley, New York, p. 429.

McCurley, M.F., G.J. Bayer and S.A. Glazer (1996). A fluorescence assay for dissolved oxygen using sol–gel encapsulated muglobin and an analogy to the inner filter effect. *Sensors Actuators B*, **35–36**, 491–496.

Mosbach, K. (1988). *Methods in Enzymology*, Vol. 137. Academic Press, New York.

Narang, U., P.N. Prasad, F.V. Bright, K. Ramanathan, N.D. Kumar, B.D. Malhotra, M.N. Kamalasanan and S. Chandra (1994a). Glucose biosensor based on sol–gel derived platform. *Anal. Chem.*, **66**, 3139–3144.

Narang, U., P.N. Prasad, F.V. Bright, A. Kumar, N.D. Kumar, B.D. Malhotra, M.N. Kamalasanan and S. Chandra (1994b). *Chem. Mater.*, **6**, 1596–1598.

NMRC report (1999). A long lifetime lactate oxidase enzyme biosensor. http://www.nmrc.ie/research/ict/~report.

Ogura, K., K. Nakaoka, M. Nakayama, M. Kobayashi and A. Fujii (1999). Thermogravimetry mass-spectrometry of urease-immobilized sol–gel silica and the application of such a urease-modified electrode to the potentiometric determination of urea. *Anal. Chim. Acta*, **384**, 219–225.

Ostraff, R.M., D. Hopkins, A.B. Haeberli, W. Baouchi and B. Polisky (1999). Thin film biosensors for rapid visual detection of nucleic acid targets. *Clin. Chem.*, **45**(9), 1659–1664.

Pandey, P.C., S. Upadhyay and H.C. Pathak (1999a). A new glucose sensor based on encapsulated glucose oxidase within organically modified sol–gel glasses. *Sensors Actuators B*, **60**, 83–89.

Pandey, P.C., S. Upadhyay and H.C. Pathak (1999b). A new glucose biosensor based on sandwich configuration of organically modified sol–gel glass. *Electroanalysis*, **11**, 59.

Pandey, P.C., S. Upadhyay, G. Singh, R. Prakash, R.C. Srivastava and P.K. Seth (2000). A new solid-state pH sensor and its application in the construction of all-solid-state urea biosensor. *Electroanalysis*, **12**, 517–521.

Pandey, P.C., S. Upadhyay, I. Tiwari and V.S. Tripathi (2001). An organically modified silicate-based ethanol biosensor. *Anal. Biochem.*, **288**(1), 39–43.

Pankrotov, I., and O. Lev (1995). Sol–gel derived renewable-surface biosensors. *J. Electroanal. Chem.*, **393**, 35–41.

Park, T.M., E.I. Iwuoha and M.R. Smyth (1997). Development of a sol–gel enzyme inhibition-based amperometric biosensor for cyanide. *Electroanalysis*, **9**(14), 1120–1123.

Patrick, F.J. (1997). *Sci. Technol. Rev.*, pp. 14–21.

Petit-Dominguez, M.D., H. Shen, W.R. Heineman and C.J. Seliskar (1997). Electrochemical behavior of graphite electrodes modified by spin-coating with sol–gel entrapped ionomers. *Anal. Chem.*, **69**, 703–710.

Ramanathan, K., M.N. Kamalasanan, B.D. Malhotra, D.R. Pardhan and S. Chandra (1997). Immobilization and characterization of lactate dehydrogenase on TEOS derived sol–gel films. *Sol–Gel. Sci. Technol.*, **10**, 309–316.

Ramanathan, K., B.R. Jonsson and B. Danielsson (2001). Sol–gel based thermal biosensor for glucose. *Anal. Chim. Acta*, **427**(1), 1–10.

Reetz, M.T., A. Zonta and J. Simpelkamp (1995). Efficienz heterogener Biokatalysatoren durch den Einfluß von Lipasen in hydrophoben Sol–Gel-Materialien. *Angew. Chem.*, **107**, 373–376.

Sampath, S., and O. Lev (1996). Inert metal modified, composite ceramic-carbon, amperometric biosensors: renewable, controlled reactive layer. *Anal. Chem.*, **68**, 2015–2021.

Santos, E.M., S. Radin and P. Ducheyne (1999). Sol–gel derived carrier for the controlled release of proteins. *Biomaterials*, **20**(18), 1695–1700.

Scheller, F., and R.D. Schmid (Eds.) (1992). *Biosensors: Fundamentals, Technologies and Applications*. VCH, Weinheim, Germany, pp. 159–174.

Schoning, M.J., D. Tsearouchas, L. Beckers, J. Schubert, W. Zander, P. Kordos and H. Luth (1996). A highly long-term stable silicon based pH sensor fabricated by pulsed laser deposition technique. *Sensors Actuators B*, **35**, 228–233.

Schubert, U., N. Husing and A. Lorenz (1995). Hybrid inorganic–organic materials by sol–gel processing of organofunctional metal alkoxides. *Chem. Mater.*, **7**, 2010–2027.

Starodub, V.M., S.P. Dikij and S.V. Svechnikov (1996). Use of the silicon crystals photoluminescence to control immunocomplex formation. *Sensors Actuators B*, **35**, 44–47.

Sun, H.T., Z.H. Chen, W. Wlodarski and M. McCormick (1996). Silica and polyimide coated quartz crystal resonators for analysis of liquids. *Sensors Actuators B*, **35–36**, 146–153.

Tatsu, Y., K. Yamashita, M. Yamoguchi, S. Yamamura, H. Yamamoto and S. Yoshikawa (1992). Entrapment of glucose oxidase in silica gel by the sol–gel method and its application to glucose sensors. *Chem Lett.*, pp. 1615–1618.

Thust, M., M.J. Schoning, J. Vetter, P. Kordos and H. Luth (1996). A long term stable penicillin sensitive potentiometric biosensor with enzyme immobilized by heterobifunctional cross linking. *Anal. Chim. Acta*, **323**, 115–121.

Trinkel, M., W. Trettnak, F. Reininger, R. Benes, O. Leary and O.S. Wolfbeis (1996). Study of the performance of an optochemical sensor for ammonia. *Anal. Chim Acta.*, **320**, 235–243.

Tsuchida, T., M. Takasugi, K. Yoda and S. Kobayashi (1985). Application of 1-(+)-lactate electrode for clinical analysis and monitoring of tissue culture medium. *Biotechnol. Bioeng.*, **27**, 837–841.

Tuller, H.L., and R. Mlcuk (1996). Photo-assisted silicon micromachining: Opportunities for chemical sensing. *Sensors Actuators B*, **35–36**, 255–261.

Ulatowska, A., H. Podbielska, M. Lechna and B. Grzegorzewski (2001). Examination of the structure of sol–gel derived matrices for optoelectronic sensors. Opt. Mater., **17**(1–2), 169–173.

Wang, B., J. Zhang and S. Dong (2000). Silica sol–gel composite film as an encapsulation matrix for the construction of an amperometric tyrosinase-based biosensor. *Biosensors Bioelectron.*, **15**, 397–402.

Wang, J. (1999). Sol–gel material for electrochemical biosensors. *Anal. Chim. Acta*, **399**, 21–27.

Wang, K.-M., L. Jun, X.-h. Yang, F.-l. Shen and X. Wang (2000). A chemiluminescent H_2O_2 sensor based on horseradish peroxidase immobilized by sol–gel method. *Sensors Actuators B*, **65**, 239–240.

Wang, R., U. Narang, P.N. Prasad and F.V. Bright (1993). Affinity of antifluorescein antibodies encapsulated within a transparent sol–gel glass. *Anal. Chem.*, **65**, 2671–2675.

Wolfbeis, O.S., I. Oehme, N. Papkovskaya and I. Kilmant (2000). Sol–gel based glucose biosensors employing optical oxygen transducers, and a method for compensating for variable oxygen background. *Biosensors Bioelectron.*, **15**, 69–76.

Wu, S., L.M. Ellerby, J.S. Cohan, B. Dunn, M.A. El-Sayed, J.S. Valentine and J.I. Zink (1993). Bacteriorhodopsin encapsulated in transparent sol–gel glass: a new biomaterial. *Chem. Mater.*, **5**(1), 115–120.

Yao, T., and K. Takashima (1998). Amperometric biosensor with a composite membrane of sol–gel derived enzyme film and electrochemically generated poly(1,2-diaminobenzene) film. *Biosensors Bioelectron.*, **13**(1), 67–73.

Yu, P., and Y.Q. Chen (2000). The study of microdialysis glucose sensor system based on co-immobilized peroxidase with glucose-oxidase by sol–gel technique. In *Proceedings of IEEE-EMBS Asia Pacific Conference on Biomedical Engineering*, Parts 1&2, pp. 415–416.

Zauner, A., R. Bullock, X. Di and H.F. Young (1995). Brain oxygen, CO_2, pH, and temperature monitoring: evaluation in the feline brain. *Neurosurgery*, **37**, 1176–1177.

Zhishang, D., W. Wenjian and J. Chin (1993). Development of sol–gel technology. *Ceramic Soc.*, **21**, 443–450.

Chapter 5

RESEARCH AND DEVELOPMENT ON BIOSENSORS FOR FOOD ANALYSIS IN INDIA

M.S. Thakur and N.G. Karanth

OUTLINE

Advances in Biosensors
Volume 5, pages 131–160.

1. INTRODUCTION

India is a vast country, which has different regional food habits depending on the socio-cultural practices and ethnic variations. While traditionally fresh food is always preferred, due to modernization of the society and the increasing trend of traditional housewives to also become professionals and working women, food habits are changing rapidly and convenience and processed foods are replacing the conventional foods in the market. With this trend, food processing is now becoming one of the most important sectors of Indian industry. The monitoring and control of quality during manufacture is one of the essential elements of quality assurance in the processed food industry. This should be carried out at appropriate points in the process so as to provide an immediate and accurate reflection of the quality status of the product being manufactured. Analysis of food ingredients also plays an important role in ensuring the health and safety of the population, which has been a major concern of food technologists and health scientists since the last century, and necessitates accurate and rapid monitoring of essential components in the food.

Food materials, particularly fresh fruits and vegetables, can be consumed as such or can be processed for wider utilization, which causes many changes during storing, processing and transportation. During these operations the quality of food undergoes changes in terms of nutritional and safety characteristics. The processed food industries are large but have lower profit margins as compared to pharmaceutical and healthcare product industries, resulting in a tendency to invest less in expensive analytical equipment in the former. However, this trend is now changing due to consumer and regulatory demands for quality and safe food, and efficient, accurate and rapid monitoring and control has become an essential requirement for the manufacturing and marketing of food products.

Food is chemically a very complex material. People's safety and health depend on the composition of the food they consume. Food analysis has two aspects: (i) to assess quality in terms of consumer acceptance and (ii) to ensure compliance with legislation in terms of safety. Several incidents of food poisoning have been reported in the recent past in India. To keep the quality of food material at an acceptable and safe level, proper monitoring of the food ingredients and contaminants, including toxins, is essential. It is also mandatory to keep the standard of food materials as prescribed by government policies of each country in order to ensure a safe food supply. A rationale for food analysis is to ensure that food is of satisfactory quality at harvest and that this is retained until it is consumed. A number of analytical methods have been in vogue for testing the quality of food to ensure compliance with legislation. For new tests to be adopted by the food industry, a high level of reliability of the analysis to match the results of existing methods is necessary for acceptance by both industry and government.

Food samples may be analysed using a sophisticated range of laboratory equipment such as gas chromatography, mass spectrometry, spectrophotometers, high-performance liquid chromatography, etc. Conventional methods for analysis of several food ingredients are time consuming, tedious, often expensive and unspecific. They cannot be used on-line. Quite often, rapid and reliable analysis is necessary for quick decision making. Biosensors fulfill many of these requirements, and have many advantages as compared to conventional analysis. Conventional analytical methods are laborious and often involve pretreatment, and it is not possible to use them to get quick and reliable analyses. One of the modern techniques which overcomes the difficulties encountered in conventional methods of analysis is the use of biosensors. Biosensors are analytical devices consisting of a biologically active material like an enzyme, antibody or binding protein and an electronic or optical system to convert the biochemical events into quantifiable electronic signals that can be processed. An important advantage of a biosensor is its amenability for on-line monitoring of food and biotechnological processes.

2. APPLICATION OF BIOSENSORS FOR FOOD ANALYSIS

Quality control during the manufacturing process is one of the essential requirements in the food industry. This should be carried out at appropriate points in the process so as to provide an immediate and accurate reflection of the quality status of the product being manufactured. Tests should include physical, organoleptic, chemical, microbiological and toxicological characteristics. The manufacturer must ensure that no product is dispatched until all specified quality tests have been satisfactorily completed (Govinda 1995).

Increased pollution of the environment in recent years has resulted in the occurrence of contaminants in foods mainly due to several practices of modern society. One of the main sources of contamination is pesticide residues from agricultural practices, industrial and domestic wastes. Food contamination may be classified into three groups: (A) Biological: aflatoxin, coliforms, *Salmonella*, and *Staphylococcus*. B) Chemical: organochlorine and organophosphorous pesticides, biphenyls, nitrosamines and tricresyl phosphates. C) Heavy metals like lead, cadmium, mercury, and arsenic. Most of these contaminants are highly toxic and some are cumulative in their effect and may cause serious health hazards. Outbreaks of food poisoning cases caused by the above-mentioned contaminants have been reported. Assessment of the levels of these contaminants has become vital and can be achieved through monitoring using modern techniques (Nagaraja 1995).

Monitoring of food ingredients using biosensors is a recent development and is finding increasing application in food processing industries worldwide. There are six stages of biosensor development for food analysis (Scott 1998):

(1) Identify the target analyte and target foods.
(2) Identify the critical user issues.
(3) Develop the analyte recognition and transduction system.
(4) Overcome problems associated with the target food matrices.
(5) Address the critical user issues in format design.
(6) Refine the commercial aspects of the instrument.

There are over 2500 commercially available enzymes that can be used for biosensor development (Turner *et al.* 1989). Amperometry was the basis of the first biosensor, described by Clark and Lyons (1962) for the analysis of glucose. Some potential uses for these biosensors are summarized in Table 1.

Table 1. Potential compounds which can be monitored using biosensors

Group of compounds	Analyte
Amino acids	alanine, arginine, asparagine, aspartic acid, cystein, glutamine, glutamic acid, glutathione, histidine, leucine, lysine, methionine, *N*-acetylmethionine, phenylalanine, sarcosine, serine, tyrosine, tryptophan, valine
Amines, amides, heterocyclic compounds	aminopyrine, aniline, aromatic amines, acetocholine, choline, phosphatidylcholine, creatine, uric acid, xanthine, hypoxanthine
Carbohydrates	amygdalin, galactose, glucose, glucose-6-phosphate, lactose, maltose, sucrose, starch
Carboxylic acids	acetic acid, formic acid, gluconic acid, iso citric acid, ascorbic acid, lactic acid, malic acid, oxalic acid, pyruvic acid, succinic acid, nitriloacetic acid
Gases	NH_3, H_2, CH_4, SO_2, NO
Cofactors	AMP, ATP, NAD(P)H, H_2O_2
Inorganic acids	fluoride, nitrite, nitrate, phosphate, sulphate, sulphite
Heavy metals	Hg^{2+}, Zn^{2+}
Complex compounds	Antibiotics, carbohydrates, assimilable substances, mutagens, vitamins
Alcohols, phenols	acetaldehyde, bilirubin, catechol, cholesterol, ester, ethanol, glycerol esters, methanol, phenol

3. BIOSENSORS FOR CARBOHYDRATES

Carbohydrates form one of the most abundant organic components in nature. They can be defined as a class of compounds containing both simple sugars as well as polymers built up by simple sugars. Carbohydrates are the primary component of any food, largely required for the growth of microorganisms, plants and animals. Quality and nutritional value of food largely depend on its carbohydrate composition. Many biotechnological processes for cultivation of microorganisms for the production of valuable chemicals require carbohydrates as an essential ingredient of the medium for fermentation.

3.1. Starch

Starch is the energy reserve source of plants as well as living beings on the earth. The determination of starch and glucose is a tedious process using conventional analytical methods. The various steps are time consuming, involving the addition of hydrolyzing chemicals and incubation followed by glucose estimation. In a flow injection system with a multienzyme-based biosensor described by Emneus and Gorton (1990), the following detection principle was used:

$$\text{starch} \xrightarrow{\alpha\text{-amylase}} \alpha\text{-limit dextrins} + \text{maltose},$$

$$\text{starch} + (\alpha\text{-limit dextrins}) \xrightarrow{\text{amyloglucosidase}} \alpha\text{-D-glucose} + \beta\text{-D-glucose},$$

$$\beta\text{-D-glucose} + O_2 + H_2O \xrightarrow{\text{glucose oxidase}} \text{D-gluconate} + H^+ + H_2O_2.$$

Amperometry can be used for the detection of the product; hydrogen peroxide can be correlated with starch contents.

3.2. Sugar Analysis

In the food industry, estimation of sugars is frequently encountered. In sugar manufacture, beverages and fruit juice industries the concentration of sugars needs to be estimated for effective quality control. Based on immobilized single- and multienzyme systems and amperometric detection, a suitable biosensor for the monitoring of glucose and sucrose in food and fermentation processes has been developed at CFTRI, Mysore, with the following industrial applications:

(1) Sugar manufacture.
(2) Confectioneries industry.
(3) Malting and brewing.
(4) Glucose and glucose saline water manufacturing.
(5) Fruit juice and soft drink manufacturing.
(6) Alcohol production by fermentation.

For the measurement of glucose, immobilized glucose oxidase (GOD) is used. In the presence of glucose oxidase enzyme, glucose undergoes oxidation according to the following reaction:

$$\beta\text{-D-glucose} + O_2 \xrightarrow{\text{GOD}} \text{gluconic acid} + H_2O_2.$$

In terms of electron transfer, the reaction can be written as

$$O_2 + 4H^+ + 4e^- \rightarrow 2H_2O.$$

The depletion of oxygen at the electrode caused by the biochemical reaction also involves consumption of electrons, resulting in an

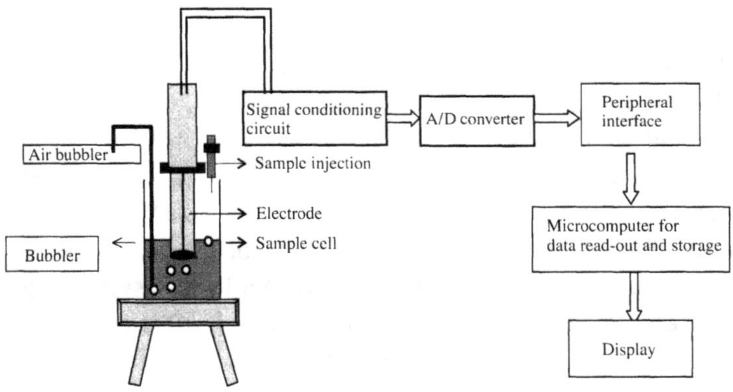

Figure 1. Schematic diagram of CFTRI Biosensor.

electrochemical signal that is proportional to the concentration of glucose in the sample. This signal is conditioned, amplified and monitored through an amperometric detector system.

For sucrose, a disaccharide, the following multienzyme system consisting of invertase, mutarotase and GOD was used, leading to the following sequence of biochemical reactions:

$$\text{sucrose} + H_2O \xrightarrow{\text{invertase}} \alpha\text{-D-glucose} + \text{fructose},$$

$$\alpha\text{-D-glucose} \xrightarrow{\text{mutarotase}} \beta\text{-D-glucose},$$

$$\beta\text{-D-glucose} + O_2 \xrightarrow{\text{GOD}} \text{gluconic acid} + H_2O_2.$$

Sucrose is first converted into α-D-glucose and fructose by invertase. Due to the stereospecificity, α-D-glucose is not acted upon by GOD and needs to be converted into β-D-glucose by using mutarotase, which is then further oxidized to gluconic acid and hydrogen peroxide. The electronic signal generated will now be proportional to the concentration of sucrose in the sample.

A microprocessor-based biosensor for glucose and sucrose was developed at CFTRI, Mysore using immobilized enzymes and an amperometric detector system, schematically shown in Figure 1. The system uses a microprocessor with alphanumeric display, RS-232 port and recorder output. Immobilized glucose oxidase was used for the glucose biosensor and an immobilized multienzyme system containing glucose oxidase, invertase and mutarotase for the sucrose biosensor. Immobilization of enzymes was carried out using various methodologies, and the glutaraldehyde cross-linking method was found to be most suitable. Calibration was carried out with known concentrations of glucose and sucrose which gave excellent linearity and reproducibility in the concentration range 1–10% for both glucose and sucrose, which is suitable for the food and fermentation industries.

Optimization of the multienzyme system for the sucrose biosensor was carried out using response surface methodology, using a three-factor, three-variable experiment design (Gouda *et al.* 2001a). In the range of enzyme concentration studied, the optimal concentrations of the enzymes were: invertase 10 IU, mutarotase 40 IU and GOD 9 IU, giving a biosensor response time of 2–3 minutes.

3.3. Stability of Immobilized Enzyme used in Biosensor

An important consideration for the application of enzyme-based biosensors is stability with temperature and pH and the operational stability of the enzyme sensing element. Immobilization of enzymes by glutaraldehyde cross-linking is a well-known method. However, it is known that glutaraldehyde, being a strong biofunctional reagent, modifies the enzyme drastically, leading to conformational changes and loss of activity. This deleterious effect can be minimized by using inert proteins like BSA, gelatin, thrombin and lysine, which act by reducing the intramolecular cross-linkage within the enzyme and enhancing the intermolecular linkage between the enzyme and the inert protein. It has been reported (Chang and Mahoney 1995) that a complementary surface protein can play an important role in stabilization of the enzyme. Therefore, it is possible that other stable proteins showing better complementarity with the enzyme provide better stability of the enzyme. This protein may be catalytically active or inactive but if it is an enzyme, its product should not interfere with the biochemical reaction at the desired enzyme electrode surface.

Operational life of the biosensor element is practically important, particularly while monitoring food and fermentation processes, where the substrate concentration is relatively high. The activity loss due to denaturation and deactivation of the immobilized enzyme diminishes the life of the biosensor. Therefore techniques to enhance the storage and operational stability of the enzyme electrode are important in the application of electrochemical biosensors.

Operational stability studies carried out with the biosensor for glucose and sucrose indicated that the immobilized enzyme system was not very stable with commonly used stabilizing matrices such as gelatin and BSA. It was found that the lysozyme incorporated during immobilization was a superior stabilizing agent for the immobilized single- and multienzyme system as compared to gelatin and BSA for long-term usage of the biosensor at room temperature (Gouda *et al.* 2002a). Using this technique, an increased number of repeated analyses of 750 samples during 230 days for glucose and 400 samples during 40 days of operation for sucrose have been achieved. The increased operational stability of the immobilized single- and multienzyme system will improve operating cost-effectiveness of the biosensor.

Studies on thermal stabilization of single and multienzyme systems were also carried out and it was found that lysozyme can stabilize the enzymes considerably for biosensor application (Gouda *et al.* 2001b). Exposing the immobilized preparation to varying temperature, pH and denaturants revealed that lysozyme stabilizes the single- and multienzyme systems better than the other protein-based stabilizing agents such as BSA and gelatin. Ionic interaction between GOD and lysozyme was shown to play a crucial role in the enzyme stabilization, and the result also indicated that the stability of enzymes immobilized by cross-linking can be enhanced by selecting a suitable protein-based stabilizing agent (PBSA) which provides better complementarity for the desired enzyme, thus leading to favorable interaction between the enzyme and the PBSA.

One of the problems encountered in the use of a sucrose biosensor is the presence of glucose. Attempts were made at CFTRI, Mysore to analyse sucrose in the presence of glucose using a two-electrode system. A theoretical formula was derived to calculate the sucrose concentration in the presence of different concentrations of glucose in the sample by making use of the calibration graphs for pure glucose and sucrose.

3.4. Biosensor as a Tool to Study Enzyme Denaturation–Renaturation

Denaturation of immobilized enzymes is a critical problem in industrial applications. The reversal of the denaturation in order to regain the original activity is important from the application point of view. The ability to regain the functional state of the immobilized enzyme after undergoing denaturation is adversely affected by a combination of intermolecular and intramolecular interactions responsible for protein folding (Kauzamann 1959). Most of the renaturation studies reported have been carried out on a single protein immobilized on solid support. Trivedi *et al.* (1997) demonstrated the importance of heteromolecular interaction between the acidic and basic proteins in the renaturation of lysozyme (basic protein) in solution, which according to them, was drastically affected due to the electrostatic interactions and aggregation by the addition of acidic proteins like BSA and alcohol dehydrogenate (ADH) to the renaturation buffer.

While it is known that the electrostatic interactions between proteins which play a crucial role in the renaturation and aggregation of proteins can be avoided by immobilization, information on the renaturation of immobilized multiple proteins is not available. Also, while it is known that proteins like lysozyme can by themselves exhibit reversible denaturation with denaturants like GdmCl and urea (Goldberg *et al.* 1991 and Perrauddin *et al.* 1976), it has not been known whether their incorporation during enzyme immobilization might help the process of renaturation of the desired enzymes. In our laboratory we had earlier observed that incorporation of lysozyme during the immobilization step increases the thermal stability of GOD, which may be attributed to the ionic interactions between lysozyme (basic protein) and GOD (acidic protein) (Gouda *et al.* 2001b). In this context, we investigated the denaturation–renaturation behavior of immobilized GOD, and the role of ionic interactions and influence of refolding ability of the stabilizing agent on the enzyme, on which no information is available.

Denaturation–renaturation behavior of proteins can be studied by using CD or NMR measurements. However these methods have several constraints with respect to their application to immobilized enzymes. As an alternative approach the activity of enzymes like GOD in immobilized form can be easily followed with the help of a biosensor employing a dissolved oxygen electrode wherein the current signal for a fixed glucose concentration bears a definite relationship with the activity of the immobilized enzyme. Using this principle, GOD immobilized with various stabilizing agents like BSA, gelatin, lysozyme and polyethylene-imine (PEI) with glutaraldehyde cross-linking was investigated by us for the denaturation–renaturation behaviour with respect to enzyme activity measured by the biosensor (Gouda *et al.* 2002b).

Denaturation was carried out for one hour at various concentrations of guanidine hydrochloride (GdmCl) in 50 mM phosphate buffer, pH 6.0 at 25±1°C. The results show that lysozyme incorporation during enzyme immobilization was very effective, giving a 72% original activity within 20 minutes of renaturation. After five cycles of repeated denaturation and renaturation with 8 M GdmCl, GOD immobilized with lysozyme still retained 70% of the original activity. These results indicate that the refolding ability of lysozyme, glutaraldehyde cross-linkages between lysozyme and GOD, together with ionic interactions between them play an important role in the denaturation–renaturation behavior of the immobilized enzyme. Further, they suggest that the refolding ability of stabilizing agents can modulate the reversible denaturation behavior of the immobilized enzyme. These results should provide useful information in understanding the influence of the refolding ability of the stabilizing agent on the immobilized enzyme. Incorporation of lysozyme during immobilization of enzymes can be employed as a useful tool for the intrinsic evaluation of the various refolding reagents by avoiding aggregation, which is a common problem in the denaturation study of soluble protein. By selecting a suitable stabilizing agent which has a refolding ability, the desired enzyme can be efficiently renatured (Gouda *et al.* 2002b).

3.4.1. Construction of a Prototype Instrument for Industrial Application

With a view to commercialization of the instrument, collaboration was established with an instrument manufacturing company. A prototype biosensor instrument was constructed (Figure 2) and has undergone tests and troubleshooting in the laboratory for glucose and sucrose analyses. Later it was field-tested in the sugar factories and confectionery industries.

The biosensor instrument developed (Figure 1) for the measurement of glucose and sucrose in food and fermentation samples has the following specifications:

- Detection range of glucose and sucrose: 1–10%.
- Accuracy: ±3%.
- Detection time: 3–4 minutes.
- Life of enzyme membrane: over 60 days of operation
- Working temperature: 27–32°C.

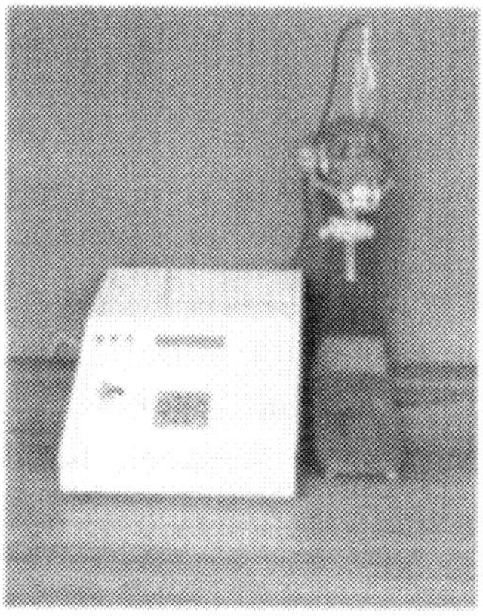

Figure 2. CFTRI Biosensor for sugar analysis.

3.5. Application of Biosensors for the Sugar Industry

Rapid and reliable methods of analysis can be most useful for better process control and improved plant performance in taking timely decisions in the various steps of sugar manufacture such as assessment of sugar content in

- sugar cane,
- primary juices,
- mixed juices,
- clear juices,
- massecuite of various stages, and
- molasses.

The present method for the analysis of sugars adopted in the industry is time consuming and not specific. Quick analysis will be advantageous for good quality sugar grains. Feeding of the sugar cane juice from one reaction tank to another reaction tank is dependent on the sugar composition in the stream. Therefore quick analysis is extremely important.

It is also important to harvest the sugar cane crop at the highest sugar content. Again, a biosensor may be useful here. A biosensor could be used to determine the sugar content in sugar cane quickly and conveniently. Based on the data obtained by analysis, sugar cane can be harvested at the maximum sugar content to enable maximum efficiency.

4. BIOSENSORS FOR ORGANIC ACIDS

4.1. L-Lactate

Analysis of lactic acid is important in dental care, food processing, fermentation and plastic manufacturing. In food processing lactic acid is essential for the manufacturing of cheese, pickles, dahi, yoghurt and fermented meat. Also, D-lactate monitoring is essential in vacuum-packed chilled meat. The L-form lactic acid widely occurs in food, fermentation and blood samples. The presence of L-lactic acid in food is also taken as an index of spoilage, contamination by microorganisms, and the quality of food in general.

4.1.1. Development of Biosensor for L-Lactate at CFTRI, Mysore

Enzyme-based biosensors for the determination of lactic acid in milk and other fermented milk products have been reported (Macini *et al.* 1984, Mulchandani *et al.* 1995). However, these are based on lactate oxidase (LOD) where the possibility of interference from other oxidizing and reducing substances prominently exists. Enzyme electrodes for lactic acid based on lactate dehydrogenase (LDH) have also been reported but exhibit reduced stability due to fouling of electrodes caused by oxidation of products of reduced nicotinamide adenine dinucleotide (NADH) or mediators (Scheller *et al.* 1987). Further, they are expensive due to the requirement of NAD (nicotinamide adenine dinucleotide), an expensive and unstable co-substrate (Kulys *et al.* 1993).

LOD-based biosensors have the disadvantage of enzyme inhibition of hydrogen peroxide (H_2O_2) and also nonspecific electro-oxidation of compounds such as ascorbic acid, uric acid, glutathione etc., at the relatively high potential required for electrochemical oxidation of H_2O_2. The presence of reducing substances such as ascorbic acid poses a problem in the detection for H_2O_2, and these interfere in the analysis.

The enzyme lactate mono-oxygenase (LMO) (EC 1,13,12,4) which converts lactate to acetate, carbon dioxide and water without the production of hydrogen peroxide has been used at CFTRI, Mysore, thus overcoming the problems with LDH- and LOD-based biosensors outlined above. Since no production or detection of H_2O_2 is involved, there is neither the inactivation of the enzyme nor interference from compounds like ascorbic acid. LMO neither requires the cosubstrate NADH nor produces H_2O_2.

4.1.2. Construction of an LMO Enzyme Electrode

The principle on which the biosensor device is based is the electrochemical changes brought about by the biochemical reactions catalysed by the enzyme LMO in the vicinity of the sensing element of a Clark dissolved oxygen electrode, resulting in a changing current. This current, when suitably conditioned, amplified and detected through a signal-handling system, gives a voltage which is proportional to the L-lactate content in the sample in which the biosensor element is placed. The biochemical reaction taking place is

$$\text{L-lactate} + O_2 \xrightarrow{\text{LMO}} \text{acetate} + CO_2 + H_2O.$$

The L-lactate in the sample is oxidized to acetate catalysed by LMO, resulting in the consumption of oxygen. This oxygen consumption, at an applied potential of –650 mV, is accompanied by acceptance of electrons resulting in the following cathodic amperometric reaction:

$$O_2 + H_2O + 4e^- \rightarrow 4OH^-.$$

The consumption of electrons decreases the current and this is related to the concentration of L-lactate in the sample in a quantitative manner. Based on the above principle, a batch-type L-lactate biosensor for analysis in the concentration range 50–800 mg/dL has been constructed at CFTRI, Mysore and the technology transferred to industry. This feature has enhanced operating life of 60 days for enzyme sensing element of the biosensor which is covered by patent.

4.2. Ascorbic Acid

Ascorbic acid is an important vitamin for humans and essential for many physiological functions. It occurs naturally in many foodstuffs (fruits, vegetables, dairy products, meat, etc.) and is frequently added to processed foods as an antioxidant. Ascorbic acid is a very effective synergist for phenolic antioxidants such as propylgallate, hydroquinone, catechol and nor-dihydroguaiaretic (NDGA) acid where they are used to inhibit oxidative rancidity in fats, oil and other lipids of long-chain fatty acids. Esters of ascorbic acid are more soluble and suitable for use with lipids than ascorbic acid. Determination of ascorbate levels in these matrices is relevant since they are an indicator of freshness (Thakur and Karanth 1999). A wide variety of cells and tissues of plants and animals can be used for detection of several analytes. Tissues can be used as the analytical component for biosensors that functionally respond to biological, chemical, or physical stimuli which can be interfaced with an optoelectronic system for the quantification of the analyte.

Work has been carried out at CFTRI, Mysore on the development of a tissue-based biosensor for L-ascorbic acid analysis in food and pharmaceutical samples. For this, immobilized cucumber tissue sandwiched between two semipermeable membranes and an amperometry-based biosensor system has been used for the analysis of samples containing L-ascorbic acid. It was shown that the performance of the biosensor developed was comparable to HPLC method. About 60 repeated analyses could be carried out using a single membrane containing the cucumber tissue. Use of enzyme-rich plant tissue instead of purified enzyme eliminates the need for extensive and expensive protein purification steps and enzyme immobilization needed in case of pure enzyme.

The biochemical reaction occurring at the electrode surface is given as:

$$\text{L-ascorbic acid} + O_2 \xrightarrow{\text{ascorbic acid oxidase}} \text{dehydroascorbate} + H_2O.$$

The reaction results in the consumption of oxygen from the solution surrounding the electrode surface, causing a decrease in the voltage response which is proportional to the ascorbic acid concentration, and monitored by the detector system.

5. BIOSENSORS FOR ORGANOPHOSPHOROUS PESTICIDES

A variety of pesticides and herbicides have been used extensively in agricultural practice to increase productivity, leading to pesticide residues in soil, water and food, particularly in developing countries like India. These residues are capable of causing severe health hazards even at trace levels. Therefore, it is essential to monitor the levels of these pesticides in food and the environment. Some of the chemical methods used for pesticide analysis include HPLC, GC, ELISA and spectrophotometric methods, which are laborious and expensive.

Discontinuation of the use of organochlorine pesticides in most countries has led to the use of other classes of pesticides such as organophosphates (OP) and carbamates. The solubility of OP's in water is higher than that of organochlorides, increasing their environmental mobility in the soil and thereby making them ubiquitous. Exposure to OP even at trace levels poses health hazards as OP's affect neurotransmission. Therefore efficient and reliable monitoring of OP pesticides is very important. Known chemical methods for OP analysis have several disadvantages. An alternative modern technique is the use of biosensors.

5.1. Detection of OP Pesticides using Acetyl Cholinesterase

Single- as well as multienzyme-based systems for OP pesticide detection have been reported. Single-enzyme-based biosensors use either acetylcholinesterase (AChE) or butyrylcholinesterase (BuChE) as the biological component, and thiocholine production is monitored amperometrically (Hartley and Hart 1994) or acid production is monitored potentiometrically. Multienzyme-based biosensors use cholinesterase in conjunction with choline oxidase and measure hydrogen peroxide production (Rekha *et al.* 2000b) or oxygen consumption (Gouda *et al.* 2001a,b). There are also reports of using acid phosphatase (Gouda *et al.* 1997) and alkaline phosphatase (Ayyagari *et al.* 1995) for OP pesticide determination.

Multienzyme-based biosensors for OP pesticide detection couple an enzyme which is inhibited by the pesticide in conjunction with an indicator enzyme which uses the product of the first enzymatic reaction as the substrate with concomitant consumption of oxygen which can be readily monitored using a Clark oxygen electrode. Even though the pesticide inhibits only one of the two enzymes, monitoring of this inhibition is not possible without coupling it with a second enzymatic reaction. This requires both enzymes to be catalytically active. The second enzyme in general does not have the same optimum working conditions like pH, temperature, molarity, etc., necessitating a compromise while selecting the operating parameters. This results in both the enzymes functioning at a sub-optimal level. This disadvantage can be overcome by using single-enzyme systems like the one described by Rekha *et al.* (2000a,b). The use of an oxidase enzyme which is readily inhibited by the pesticide avoids the need for coupling a second enzymatic reaction merely for the purpose of monitoring the inhibition reaction. Considerable research has been carried out at CFTRI, Mysore, on the development of single- and multienzyme-based amperometric biosensors for OP pesticide detection (Gulla *et al.* 2001). It is known that organophosphates exhibit their pesticidal power through a strong inhibition of acetylcholinesterase (AChE) activity. This inhibition principle has been used to develop a biosensor for the detection of OP pesticides.

The principle involved in an AChE-based biosensor is the hydrolysis of acetylthiocholine to thiocholine and acetic acid catalyzed by AChE as follows:

$$H_2O + (CH_3)_3{-}N{-}(CH_2)_2{-}S{-}\underset{\displaystyle CH_3}{\underset{|}{C}}{=}O$$

$$\downarrow \text{AChE}$$

$$(CH_3)_3{-}N{-}(CH_2)_2{-}SH + CH_3{-}COOH$$

$$\downarrow +410\,\text{mV anodic oxidation}$$

$$\begin{array}{l} S{-}(CH_2)_2{-}N{-}(CH_3)_3 \\ | \\ S{-}(CH_2)_2{-}N{-}(CH_3)_3 + 2H^+ + 2e^-. \end{array}$$

Thiocholine oxidises at the electrode surface at +410 mV, which increases current output. Organophosphorous pesticide inhibits AChE, decreasing thiocholine production and in turn causing a decrease in current output, which is correlated with the pesticide concentration.

Using this principle, a laboratory biosensor has been constructed at CFTRI, Mysore, for paraoxon with a sensitivity of 0.5 ppm. This sensitivity is not quite adequate for practical applications and efforts are in progress to improve the biosensor performance.

5.1.1. Enzyme Deactivation Problem with AChE

While biosensors based on AChE inhibition have been known for monitoring of OP pesticides, strong inhibition of the enzyme is a major drawback in the practical application of the biosensor. This can be overcome at least partially by reactivation of the enzyme for repeated use. In our laboratory enzyme reactivation by oximes was explored for this purpose (Gulla *et al.* 2002). Two oximes viz., 1,1′-trimethylene-bis-4-formylpyridinium bromide dioxime (TMB-4) and pyridine 2-aldoxime methiodide (2-PAM) were compared for the reactivation of the immobilized AChE. TMB-4 was found to be a more efficient reactivator under repeated use, retaining more than 60% of initial activity after 11 reuses, whereas for 2-PAM the activity retention dropped to less than 50% after only 6 reuses. Investigations also showed that reactivation must be effected within 10 minutes after each analysis in order to eliminate the aging effect, which reduces the efficiency of reactivation.

5.2. Detection Based on Acid Phosphatase Inhibition

An amperometry-based biosensor has also been developed to analyse the OP pesticide using the dual enzyme acid phosphatase and glucose oxidase (GOD) (Gouda *et al.* 1997).

The biochemical reactions occurring in this dual enzyme system are

$$\text{glucose 6-phosphate} + H_2O \xrightarrow{\text{acid phosphatase}} \text{glucose} + HPO_4^3,$$

$$\text{glucose} + O_2 \xrightarrow{\text{GOD}} \text{gluconic acid} + H_2O_2.$$

Glucose 6-phosphate is converted to glucose in the presence of acid phosphatase and further oxidized to gluconic acid catalyzed by GOD. Activity of the acid phosphatase is inhibited by the OP pesticide, resulting in a reduction in the second reaction, monitored in terms of a change in output voltage of the biosensor, which is proportional to the pesticide concentration.

Using the above system, an amperometric biosensor consisting of a potato layer rich in acid phosphatase and an immobilized GOD membrane, when operated in conjunction with a Clark-type electrode, detected the pesticide. A notable advantage of this biosensor is that the inhibition of acid phosphatase by the pesticide is reversible and thereby eliminates the serious problem of enzyme inactivation and the necessity for its reactivation, which is not efficient.

5.3. Biosensor Based on Ascorbate Oxidase

Biochemical studies have indicated that a number of enzymes including ascorbate oxidase are inhibited by OP pesticides (Rekha *et al.* 2000a).

A biosensor for the quantification of OP pesticides by making use of the inhibition of ascorbic acid oxidase enzyme has been developed at CFTRI, Mysore (Rekha *et al.* 2000a). In this sensor, cucumber (*Cucumis sativus*) tissue, rich in ascorbic acid oxidase, was used for the detection of OP pesticide ethyl paraoxon, which inhibited the ascorbic acid oxidase activity. The optimal concentration of ascorbic acid used as substrate was found to be 5.67 mM. The biosensor response was found to reach steady state within 2 minutes. A measurable inhibition (>10%) was obtained with a 10 min incubation of the enzyme electrode with different concentrations of the pesticide. A linear relationship was obtained between the percentage inhibition of the enzyme substrate reaction and the pesticide (ethyl paraoxon) concentration in the range 1–10 ppm.

The biochemical oxidation of L-ascorbate catalysed by ascorbate oxidase is given by

$$2\,\text{L-ascorbate} + O_2 \xrightarrow{\text{ascorbate oxidase}} 2\,\text{dehydroascorbate} + 2H_2O.$$

Ascorbic acid is oxidized to dehydroascorbate and water in the presence of oxygen. When this reaction is made to occur on the sensing element of a conventional 'Clark' dissolved oxygen electrode, there is a local oxygen depletion, causing the current response of the electrode to decrease. The enzyme activity is inhibited by organophosphorous pesticides, resulting in a reduction in the biochemical reaction, monitored in terms of a change in response of the biosensor proportional to the pesticide concentration.

5.4. Immunosensors for Pesticide Analysis

Techniques based on the immune response principle involving antigen–antibody reactions and spectrophotometric or surface plasmon resonance (SPR) detection, can give accurate and sensitive detection of pesticides. Work is in progress at CFTRI, Mysore, on the "development of immuno-bioreactor-based biosensors for detection of pesticides and herbicides in water samples" (Gulla *et al.* 2001). The work involves raising of antibodies against organophosphorous pesticides (ethyl parathion and methyl parathion) and immobilization on suitable carrier to develop a bioreactor for flow injection studies. Tagging of enzymes, fluorescent markers like FITC to antibodies or antigens are under progress for the development of immunosensors with high sensitivity.

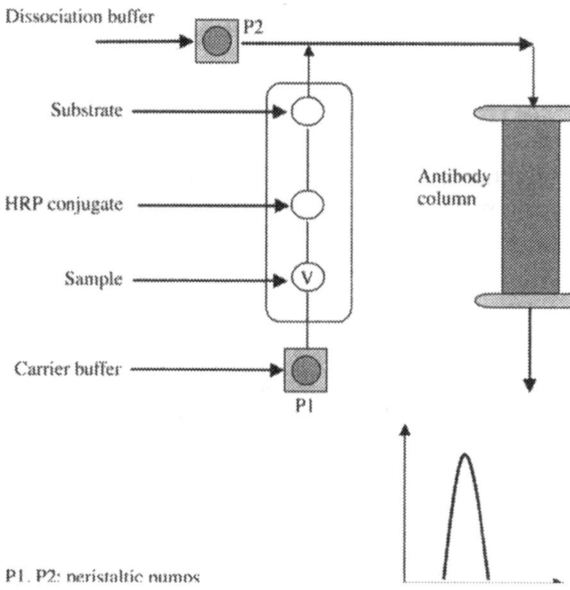

Figure 3. Schematic diagram of Flow Injection Immunosensor.

Polyclonal antibodies against organophosphorous pesticides (ethyl parathion and methyl parathion) have been raised in rabbits. An antibody column was prepared by immobilizing the antibody on controlled pore glass by silanization and glutaraldehyde cross-linking (Cass and Ligler 1998). The FIA set-up, and the sequence of events during the assay, are shown in Figures 3 and 4, respectively. The antibody column was connected to the flow injection system using two peristaltic pumps, three two-way valves and T-loops. The flow starts from the carrier buffer reservoir, and ends at the antibody column. In order to pass the analyte, HRP-conjugate and substrate along with carrier buffer loops and manually operated two-way valves were used which were connected to the main tubing as shown in Figure 3. The carrier buffer was continuously flowing during all the operations except when the regeneration buffer had to be passed. At any given time, only one valve is kept open. After each valve operation, buffer is passed to ensure removal of the excess reagent in the tubing and column. With the FIA set-up constructed as described above, a linear calibration was obtained for DDT with which analysis of unknown samples could be done; the results were comparable to GC and ELISA. After each analysis, in order to reuse the column, regeneration was attempted by passing the Gly-HCl buffer (pH:2.5). Investigations are

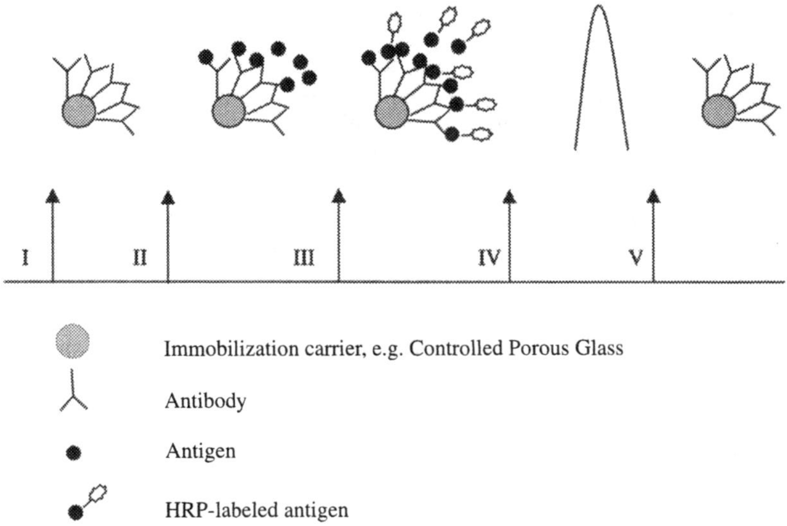

Sequential events:

I Antibody immobilized on CPG

II Pesticide sample injected and bound to the antibody

III HRP-labeled pesticide injected

IV Substrate for peroxidase injected

V Glycine-HCl buffer (pH:2.5) (for dissociation) passed for reconditioning the antibody column

Figure 4. Schematic diagram of sequential events in Flow Injection Immunosensor for pesticide.

planned using a FIA system for the analysis of OP pesticides like ethyl parathion (folidol) and methyl parathion (metacid).

6. BIOSENSORS FOR DETECTION OF TOXINS AND PATHOGENS

Food-related illnesses caused by food-borne pathogens and microbial toxins are common in India. Toxic metabolites produced by fungal growth pose a serious health threat to the population. Important toxins produced by fungi are aflatoxin B_1, fumonisin B_1, ochratoxin A,

Table 2. Pathogenic microorganisms and toxins affecting food safety[a]

Toxins/pathogens	Food samples
Clostridium botulinium neurotoxins A, B, F, E, G	sheep, cattle, meat extract, foods, stools, feces, fish, human tissue, salmon, beef, pork, canned food
Staphylococcus aureus enterotoxins A, B, C, D, E; pure toxins	milk, cheese, nonfat dry milk, hamburger, boiled eggs, pudding, custard, food extracts, minced meat, sausages
Mycotoxins	
Aflatoxin, ochratoxin	corn, wheat, peanut, butter, barley, milk, animal tissue, meat
Algal and sea food toxins	Sea foods
Salmonella, *Listeria monocytogenes*, *Escherichia coli*, *Vibrio* spp.	milk, cheese, meat food products, seafoods
Yersinia enterocolitica	milk
Campylobacter jejuni	chicken

[a] Data from Samarajeeva *et al.* (1991) and Deshpande and Platshon (1993).

patulin, trichothecene, citrinin, fumonisin, T_2 toxin, Zearalenon, toxins from *Clostridium perferigens*, campylobacteriosis and toxico-infections (*E. coli*); monitoring or detection of these toxins are very important in relation to human health. Conventional analytical methods have their limitations, necessitating more effective alternative methods based on enzymes, antibodies, tissues, chemoreceptors, etc.

There are a considerable number of chemicals, toxins and pathogenic microbes which may result from contamination or food processing. Toxins may be regarded as poisonous substances produced by living organisms. Table 2 shows commonly known toxins and pathogenic microorganisms occurring in foods.

6.1. Biosensors for Aflatoxin Detection

Aspergillus flavus and the closely related subspecies parasiticus have long been recognized as major contaminants in organic and non-organic food items. *A. flavus*, a common soil fungus, can infest a wide range of agricultural products. Some *A. flavus* varieties produce aflatoxins, which are carcinogenic toxins that induce liver cancer in laboratory animals (Gourama and Bullerman 1995).

A sensitive and rapid screening method for the estimation of aflatoxin M1 (AFM1) in milk has been developed (Deboevere and Vanpeteghem 1993). Milk samples were first purified by immunoaffinity chromatography (IAC) using polyclonal antibodies raised in rabbits against AFM1-bovine serum albumin (AFM1-BSA) and coupled to an activated sepharose matrix. The elutant was analyzed in an indirect competitive streptavidin-biotin modified enzyme-linked immunosorbent assay (ELISA). Microtitre plates were coated with AFM1-BSA that competes with the analyte in the sample for binding with the biotinylated antibody. Bound biotinylated antibody was detected using a streptovidin biotinylated horseradish peroxidase complex. The limit of detection of the ELISA is $2\,ng\,kg^{-1}$. By combination of IAC and ELISA, spiked milk samples were analyzed. IAC–ELISA was a sensitive and rapid method for the estimation of AFM1 in milk. A direct competitive ELISA screening method for aflalotoxins at 20 ng/g has been reported, in which test samples of corn were extracted by blending with methanol water (8:2) (Trucksess and Stack 1994).

There is very little research work in India on the development of biosensors for toxins and pathogens. However, in view of their importance and animal health, there is a need for it and a good potential for applications exists.

7. FLOW INJECTION ANALYSIS (FIA) BIOSENSOR FOR ON-LINE MONITORING OF FOOD AND FERMENTATION PROCESSES

Recent advancements in the field of biotechnology have greatly increased the scope for new on-line sensors, especially biosensors for monitoring bioprocesses. Nutritional composition and product analysis of fermentation broth will allow the technologist to optimize the fermentation processes efficiently. While biosensors are being used for the specific analysis of several analytes in food, water and environmental samples, due to the heat sensitivity of biological components they cannot easily be used for on-line analysis in fermentation processes. Therefore, a suitable system for on-line monitoring of substrate and product is needed, where continuous analysis of many analytes for longer periods is desired.

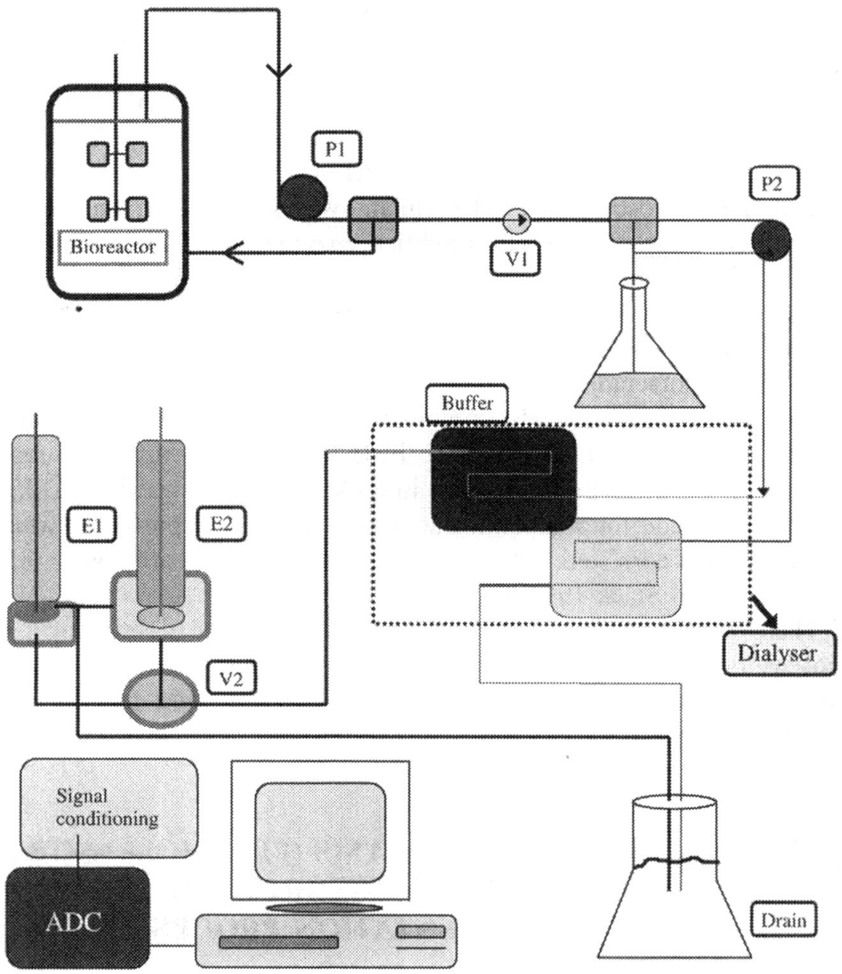

Figure 5. Flow injection analysis system for sequential analysis of glucose and L-lactate in fermentation processes. P1, P2, peristaltic pumps; V1, V2, one- and two-way valves; E1, E2, enzyme electrodes.

Flow injection analysis (FIA) is generally considered as a practicable application of biosensors for on-line monitoring in fermentation and food processing industries. FIA enables rapid, continuous analysis during fermentation avoiding contamination. Recently on-line, sequential simultaneous determination of glucose and lactic acid has been reported (Kumar *et al.* 2001); a schematic diagram of the FIA system is shown in Figure 5. Enzyme electrodes containing immobilized GOD and L-lactate oxidase were used with an amperometric detection system. A 12-bit data acquisition card with 16 analog input channels and 8 digital output channels was used. The software for data acquisition was developed using Visual C++, and was devised for sampling every hour for sequential analyses of lactate and glucose. The detection range was found to be 2–100 g L^{-1} for glucose and 1–60 g L^{-1} for L-lactate. This FIA system was used for monitoring glucose utilization and L-lactate production by immobilized cells of *Lactobacillus casei* during a lactic acid fermentation process in a recycle batch reactor. Good agreement was observed between analysis data obtained using the biosensor and data from methods of analysis of reducing sugar and L-lactate. The biosensor exhibited excellent stability during continuous operation for more than 45 days.

8. POTENTIAL OF APPLICATIONS OF FOOD BIOSENSORS IN INDIA

There are very good possibilities for the application of biosensors for food analysis in India. It is the world's largest sugar-producing country with more than 300 sugar factories. It is desirable to monitor the glucose/sucrose content during the processing of cane into sugar, and biosensors offer a rapid and accurate method superior to the conventional method. Biosensors have the potential to be used in the sugar cane field to determine the sugar content in the cane at harvest time, which can be used for pricing of the cane to the farmers. The glucose/sucrose biosensor also has a large potential for application in soft drinks and other beverage industries as well as in confectionary manufacturing industries.

India is also largest milk producer in the world. Lactic acid biosensors will find vast application in the quality control of milk, particularly in the collection of milk for processing from thousands of farmers in the villages. It will also be useful in dairy processing industries such as "dahi", yoghurt and cheese manufacture. In view of the extensive application of pesticides in Indian agricultural practice, the monitoring of pesticide residue in agricultural commodities and milk is extremely important from the point of view of the health and safety of the population. It has also become crucial for Indian food exports where stringent limits are set.

Another area of potential biosensor applications in India is during production of beverages like tea or coffee. Tea is a very popular beverage the world over and a major export earner for India. Due to stiff competition among the tea-producing countries, efficient quality monitoring and control during processing in tea manufacture has become imperative to hold its position in world market. Polyphenols and their oxidative products play a vital role in flavor and quality of tea. Currently, quality assessment is done by tea tasters, which has a inherent subjectivity. Further it does not allow monitoring during processing so as to control the process to get a consistent quality product. Therefore efficient instrumental methods are needed for this purpose. Biosensors offer an efficient, rapid and accurate method for monitoring the polyphenols and their oxidation products, the development of which is essential for the tea industry in India.

Coffee is also a major beverage crop of India with substantial consumption as well as exports. Because of the relative importance of the caffeine content in determining the quality of coffee beverages, the development of a sensitive, fast and cost-effective method for monitoring caffeine is needed. At CFTRI, work is in progress on the development of biosensors for both caffeine and polyphenols.

9. FUTURE BIOSENSOR RESEARCH

There is a strong need for the development of biosensors for detecting a large number of compounds such as aflatoxins, ochratoxins, tricothecene, pathogenic bacteria, botulinum toxins, blue/green algae toxins, pest contaminants in food and spices, and heavy metals. Biosensors for meat freshness and carcinogen detection and DNA chips/probes, electronic nose, tongue and eye are also needed for food processing applications, particularly to ensure food safety. Recently, molecular beacons have been reportedly used to detect food-borne pathogens. Molecular beacons are oligonucleotide probes that become fluorescent upon hybridization. The development of the systematic evolution of ligands by exponential (SELEX) enrichment has made possible the isolation of oligonucleotide sequences with the capacity to recognize virtually any class of target molecules with high affinity and specificity. These oligonucleotide sequences, referred to as "aptamers", are beginning to emerge as a class of molecules that rival antibodies in both therapeutic and diagnostic applications. Aptamers are different from antibodies, yet they mimic antibody properties in a variety of diagnostic formats. The demand for diagnostic assays to assist in the management of existing and emerging diseases is increasing, and aptamers could potentially fulfill molecular recognition needs in those assays.

Biosensors with receptor molecules like antibodies, DNA, DNA chips based on molecular beacons, aptamers, etc., and transducers based on fluorescence, luminescence, reflectance, optoelectronic- and electro-chemical-based techniques, will become important in the future.

Gene chips are another type of biosensor. These sophisticated silicon-chip-based genome probes (also known as DNA chips) have become available in the last few years and allow samples of DNA to be easily scanned and decoded for the purposes of sequencing, genotyping or gene expression monitoring. In 1997 gene chips contained some 400 000 DNA probes. Within a decade it will probably be possible to place a few cells in a gene-chip scanner and quickly analyze risk profile for numerous microbial diseases. Several common bacterial diseases caused by *E. coli*, *Staphylococcus aureus*, *Clostridium perfringens, C. botulinum, Campylobacter* or *Coxiella burnetii* can be easily identified using state-of-the-art technology such as molecular beacons, aptamers etc.

The detection of a specific DNA sequence is significant in many areas including clinical, environmental and food applications. The analysis of gene sequences and the study of gene polymorphisms play a fundamental role in the rapid detection of genetically modified materials, offering the possibility of performing reliable diagnosis. Methods are needed which do not require the use of labels. Biosensor technology offers such possibility with the major advantage of monitoring genetically modified organisms (GMO) hybridization in real time and with high selectivity.

10. CONCLUSION

Ensuring intake of safe food has been a major concern of food technologists and health scientists since the last century. This necessitates the proper monitoring of different components of foods and beverages along with food-borne and water-borne pathogens, toxins and pesticide residues. Rapid, reliable and accurate methods of analysis are needed to overcome the disadvantages of conventional analytical methods. Biosensors present an attractive, efficient alternative technique, and are quick and reliable in their performance. There is very good potential for the application of biosensors in India. Present research has mostly been focussed on immobilized-enzyme and amperometry-based biosensors. There is active support for biosensor research by government agencies such as the Departments of Science and Technology and Biotechnology. While the research on biosensors has reached a respectable stage, there is a need to extend the efforts in newer areas such as immunosensors, DNA probes, electronic nose, image processing, etc.

ACKNOWLEDGMENT

Financial support from the Department of Science and Technology, and the Department of Biotechnology, Government of India, as well as the Swedish Agency for Research Cooperation with Developing Countries (SAREC), Sweden, and Indo-Swiss collaboration in Biotechnology (ISCB), Switzerland, for projects on biosensor research is gratefully acknowledged.

REFERENCES

Ayyagari, M., S. Kamtekar, R. Pandey, K.A. Marx, J. Kumar, S.K. Tripathy, J. Akkara and D.L. Kaplan (1995). Chemiluminescence-based inhibition kinetics of alkaline phosphatase in the development of pesticide biosensor. *Biotechnol. Prog.*, **11**, 699–703.

Cass, T., and F.S. Ligler (Eds.). *Immobilized Biomolecules in Analysis – A practical approach*. Oxford University Press, New York.

Chang, S.B., and R.R. Mahoney (1995). Enzyme thermostabilization by bovine serum albumin. *Biotechnol. Appl. Biochem.*, **22**, 203–214.

Clark, L.C., and C. Lyons (1962). *Ann. N.Y. Acad. Sci.*, **102**, 29.

Deboevere, C., and C. Vanpeteghem (1993). Development of an immunoaffinity column and an indirect immunoassay with a biotin streptavidin detection system for Aflatoxin M(1) in milk. *Anal. Chim. Acta*, **275**(1–2), 341–345.

Deshpande, S.S., and M.C. Platshon (1993). Veterinary and environmental diagnostics. In *Diagnostics in the Year 2000*. B.P. Sharma Singh and P. Tyle (Eds.), Van Nostrand Reinhold, New York, pp. 459–525.

Emneus, J., and L. Gorton (1990). Flow system for starch determination based on consecutive enzyme steps and amperometric detection at a chemically modified electrode. *Anal. Chem.*, **62**, 263–268.

Goldberg, M.E., R. Rudolph and R. Jaeinke (1991). *Biochemistry*, **30**, 2790–2797.

Gouda, M.D., M.S. Thakur and N.G. Karanth (1997). A dual enzyme amperometric biosensor for monitoring organophosphorous pesticides. *Biotech. Tech.*, **11**(9), 653–655.

Gouda, M.D., M.S. Thakur and N.G. Karanth (2001a). Optimization of the multienzyme system for sucrose biosensor by response surface methodology. *World J. Microbiol. Biotechnol.*, **17**, 595–600.

Gouda, M.D., M.S. Thakur and N.G. Karanth (2001b). Stability studies on immobilized glucose oxidase using an amperometric biosensor – effect of protein based stabilizing agent. *Electroanalysis*, **13**, 849–855.

Gouda, M.D., M.A. Kumar, M.S. Thakur and N.G. Karanth (2002a). Enhancement of operational stability of an enzyme biosensor for glucose and sucrose using protein based stabilizing agents. *Biosensors and Bioelectron.*, **17**, 503–507.

Gouda, M.D., M.S. Thakur and N.G. Karanth (2002b). Reversible denaturation behaviour of immobilized glucose oxidase. *Appl. Biochem. Biotechnol.*, in press.

Gourama, H., and L.B. Bullerman (1995). *Aspergillus flavus* and *Aspergillus parasiticus*: Aflatoxigenic fungi of concern in foods and feeds: A review. *J. Food Protect.*, **58**(12), 1395–1404.

Govinda, A. (1995). Quality system in food industry – role of Bureau of Indian Standards. In *Trends in Food Science and Technology*, Proceedings of the Third International Food Convention, IFCON 1993. C.P. Natarajan and S. Ranganna (Eds.), pp. 999–1005.

Gulla, C., M.D. Gouda, M.S. Thakur and N.G. Karanth (2002). Reactivation of immobilized acetylcholinesterase in an amperometric biosensor for organophosphorus pesticide. *Biochim. Biophys. Chim. Acta*, **1597**, 133–139.

Gulla, K.C., M.S. Thakur and N.G. Karanth (2001). Biosensors for the detection of pesticides in food, water and environment. *BIOFUTUR* France (October), pp. 56–59.

Hartley, I.C., and J.P. Hart (1994). Amperometric measurement of organophosphate pesticides using a screen printed disposable sensor and a biosensor based on cobalt pthalocyanine. *Anal. Proc. Including Anal. Commun.*, **31**, 333–337.

Kauzamann, W. (1959). In *Advances in Protein Chemistry*, Vol. 14. C.B. Anfinsen, M.L. Anson, J.T. Edsall and F.M. Richards (Eds.), Academic Press, New York, pp. 1–63.

Kulys, J., L. Wang and A. Macsimoviens (1993). *Anal. Chim. Acta*, **274**, 53–58.

Kumar, M.A., M.S. Thakur, S. Santhuran, V. Santhuran, N.G. Karanth, R. Hatti-Kaul and B. Mattiasson (2001). An automated flow injection analysis system for on-line monitoring of glucose and L-lactate during lactic acid fermentation in a recycle bioreactor. *World J. Microbiol. Biotechnol.*, **17**(1), 23–29.

Macini, M., D. Moscone and G. Pallesche (1984). *Anal. Chim. Acta*, **157**, 45–51.

Mulchandani, A., A. Bassi and A. Nguyen (1995). *J. Food Sci.*, **60**(1), 74–77.

Nagaraja, K.V. (1995). Contaminants in foods: How to monitor and control, In *Trends in Food Science and Technology, Proceedings of the Third International Food Convention, IFCON 1993*. C.P. Natarajan and S. Ranganna (Eds.), pp. 1057–1061.

Perrauddin, J.P., T.E. Torchia and D.B. Watlaufer (1976). *J. Biol. Chem.*, **258**, 11834–11839.

Rekha, K., M.G. Gouda, M.S. Thakur and N.G. Karanth (2000a). Ascorbic acid oxidase based biosensor for organophosphorus pesticide detection. *Biosensors Bioelectron.*, **15**, 499–502.

Rekha, K., M.S. Thakur and N.G. Karanth (2000b). Biosensors for organophosphorus pesticide monitoring. *CRC Crit. Rev. Biotechnol.*, **20**(3), 213–235.

Samarajeeva, U., C.I. Wei, T.S. Huang and M.R. Marshall (1991). Appplication of immunoassay in the food industries. *CRC Rev. Food Sci. Nutr.*, **29**, 403–434.

Scheller, F.W., D. Pfeiffer, F. Schubert, R. Renneberg and D. Kirsten (1987). In *Biosensors: Fundamentals and Applications*. A.P.F. Turner, I. Karube and G. Wilson (Eds.), Oxford University Press, Oxford, pp. 315–346.

Scott, A.O. (1998). Biosensors for food analysis: perspectives. In *Biosensors for Food Analysis*. A.O. Scott (Ed.), Royal Society London, pp. 181–195.

Thakur, M.S., and N.G. Karanth (1999). Application of biosensors for global standards in food quality. In *Proceedings of the International Food Conference* (IFCON). Association of Food Scientists and Technologists, India, pp. 512–520.

Trivedi, V.D., B. Raman, C. Mohanrao and Ramakrishna (1997). *FEBS Lett.*, **418**, 363–366.

Trucksess, M.W., and M.E. Stack (1994). Enzyme-linked-immunosorbent-assay of total aflatoxin-BU, aflatoxin-B2, and aflatoxin-G1 in corn. *J. AOAC Int.*, 1060–3271, **77**(3), 655–659.

Turner, A.P.F., I. Karube and G.S. Wilson (Eds.) (1989). *Biosensors: Fundamentals and Applications*. Oxford University Press, Oxford.

Chapter 6

IMMUNOSENSORS FOR PESTICIDES MONITORING

C. Raman Suri

OUTLINE

ABSTRACT

Immunosensors are based on the binding interactions between immobilized biomolecules (antibody/antigen) on a transducer surface with the analyte of interest (antigen/antibody), resulting in a detectable signal. The sensor system takes advantage of the high selectivity provided by the molecular recognition characteristics of an antibody, which binds,

Advances in Biosensors
Volume 5, pages 161–178.

reversibly with a specific antigen. In this article various biological and transducer aspects of immunosensor development for pesticides analysis are described.

Keywords: hapten; pesticides; antibody; immunosensors

1. INTRODUCTION

The increasing use of pesticides for achieving higher agricultural yields has posed considerable problems in general health programs. These pollutants, which are highly toxic in nature, enter animals and human beings through the food chain or drinking water, and accumulate in different organs of the body affecting immune and nervous systems. The present modes of analysis of these hazardous compounds are mainly physico-chemical techniques such as gas chromatography (GC), high-pressure liquid chromatography (HPLC), thin-layer chromatography (TLC), mass spectrometry (MS), GC–MS, etc. However, most of these methods are complex, time consuming and require costly and bulky instrumentation. There has, therefore been a great demand for developing a cheap, reliable, sensitive and field-applicable assay technique for the detection and quantification of pesticides in the environment. An alternative modern technique that eliminates some of the drawbacks of existing conventional methods is the use of biosensors.

Biosensors are classified as bioelectronic products, which incorporate a biological recognition system such as enzymes, antibodies, receptors, tissues, or nucleic acid, coupled to an electronic transducer. These devices contain a reactive surface (i.e., the biochemical transducer), which generates a measurable response upon binding with an analyte. The conversion of an analyte by the biochemical transducer into another chemical species or physical properties is sensed and converted into a measurable signal by the electronic transducer (Figure 1). In practice, a combination of biomolecular immobilization and transducer technologies has been used to produce two main types of biosensors, i.e., enzyme/metabolic biosensors and bioaffinity sensors.

1.1. Enzyme (or Metabolic) Biosensors

Metabolic biosensors utilize a catalytic process to detect a substance. The reaction between an immobilized biocatalyst and the analyte results in a

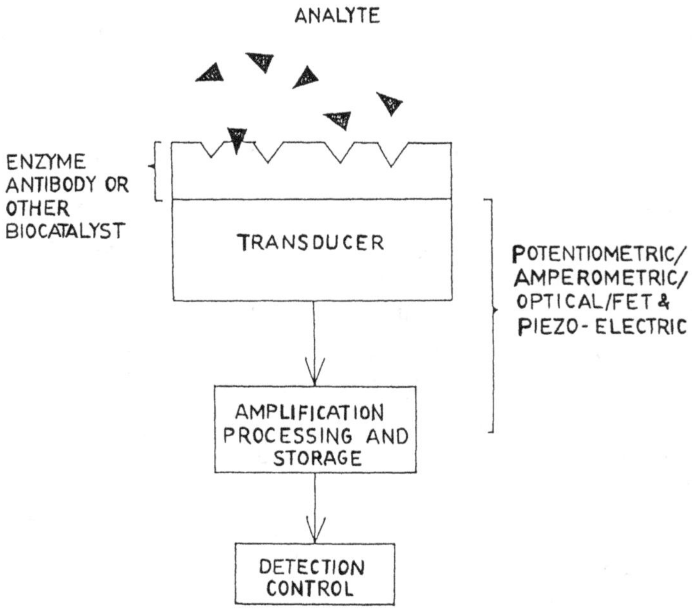

Figure 1. Basic components of biosensor design. The biospecific surface of the transducer interacts with the analyte in the sample, generating a measurable signal.

change in either concentration, such as medium oxygen or environmental parameters (pH etc.) or product formation that can be monitored by a specific electronic transducer. In the classic enzyme biosensor first described by Clark and Lyons (1962), glucose oxidase was immobilized on an oxygen electrode, and the concentration of glucose was monitored in terms of consumption of oxygen or hydrogen peroxide production. Since the time the first enzymatic bioelectrode was described, the interest in this area has grown steadily, and intense efforts are being directed towards the development of really useful and commercially viable biosensors for various analytes.

Enzyme-based biosensors have been reported for pesticide detection using cholinesterase (ChE), acetylcholinesterase (AChE), or butyryl-cholinesterase (BuChE) as the biological components (Tran-Minh *et al.* 1990, Marty *et al.* 1992, Mionetto *et al.* 1994). The basis of enzymatic biosensors for pesticide detection is the diminution of electrochemical signal by pesticide inhibition of the enzyme action. The principle involved in the AChE-based biosensor is the catalytic hydrolysis of acetylthiocholine to thiocholine, which is monitored amperometrically

(Hartley and Hart 1994) or potentiometrically (Imato and Ishibashi 1995).

1.2. Bioaffinity Sensor (Immunosensor)

Immunosensors are based on the binding interactions between immobilized biomolecules (antibody/antigen) on the transducer surface with the analyte of interest (antigen/antibody), resulting in a detectable signal. The sensor system takes advantage of the high selectivity provided by the molecular recognition characteristics of an antibody which binds, reversibly, with a specific antigen. In this approach, an antibody (Ab) interacts specifically and reversibly with its antigen (Ag) to form an immunocomplex (AbAg) according to the following equilibrium equation:

$$\mathrm{Ab} + \mathrm{Ag} \underset{K_\mathrm{d}}{\overset{K_\mathrm{a}}{\leftrightarrow}} \mathrm{AbAg},$$

where K_a and K_d are the association and dissociation rate constants. The equilibrium constant (or the affinity) of the reaction is expressed as follows:

$$K = \frac{K_\mathrm{a}}{K_\mathrm{d}} = \frac{[\mathrm{AbAg}]}{[\mathrm{Ab}][\mathrm{Ag}]}.$$

The equilibrium kinetics of antibody binding in solution suggests that both association and dissociation are relatively rapid. The direction of equilibrium depends on the overall affinity, which is basically the summation of both the attractive and repulsive noncovalent forces. An immunological complex usually shows low K_d values (in the range of 10^{-6} to 10^{-12}) and also displays higher affinity, i.e., K value (typically 10^4) of immunological system (Kabat 1976).

Antibody (or immunoglobulin) molecules react with their specific antigens via a combination of various forces such as hydrogen bonds, hydrophobic, van der Waals, Coulombic interaction forces and other repulsive forces. However, this interaction is mainly due to the entropy-driven process of hydrophobic bonding, wherein water molecules are excluded from between interacting hydrophobic surfaces (Hudson and Hay 1989). In general, affinity-based biosensors are divided into two categories: labeled and nonlabeled immunosensors.

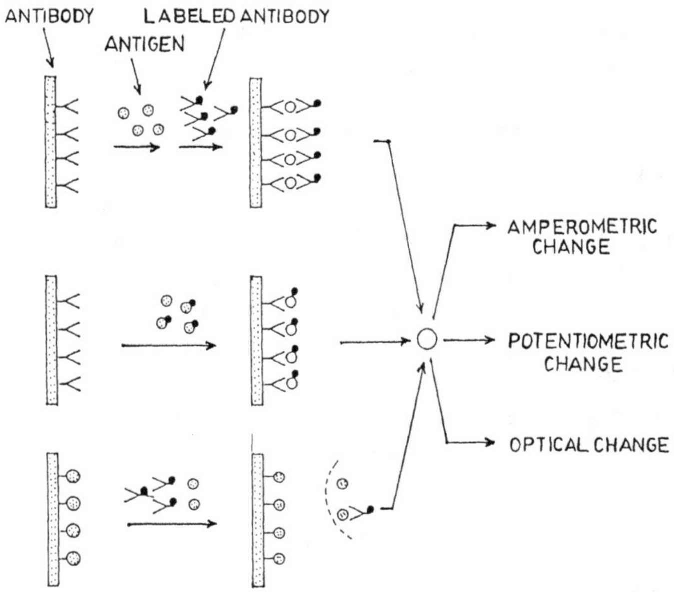

Figure 2. Principle of labeled immunobiosensors.

1.2.1. Labeled Immunosensors

In a labeled immunosensor, readily detectable labels such as radio isotopes, enzymes, fluorescent or chemiluminescent molecules are used to determine the immunocomplex. The general principle of labeled immunosensors is shown in Figure 2. An immunosensor based on the measurement of the transmembrane potential across a membrane incorporating an immobilized antibody (or antigen) was first described by Aizawa et al. (1977). The membrane-bound antibody (or antigen) specifically binds the corresponding antigen (or antibody) in solution, which results in a change in transmembrane potential. Aizawa and his group also used an enzyme linked immunoassay to construct another type of immunosensor (Aizawa et al. 1979). In this approach, an oxygen electrode was covered with an antibody binding membrane. A catalase-labeled antigen was added to a solution containing the antigen to be determined. Both antigens competitively react with the membrane-bound antibody. The labeled enzyme was then assayed by measuring the sensor output after the addition of hydrogen peroxide. For pesticide analysis by immunosensors, fluorescent dyes such as fluorescein isothiocyanate (FITC), fluorescein, or red light emitting dye, Cy5.5, are most widely used as labels. Klotz and his colleagues developed an optical immunosensor based on measurement of either fluorescence excitation or emission via the evanescent field of the waveguides, which allow real-time monitoring of the labeled antibodies (Klotz *et al.* 1998). The detection limit of the assay was in the sub-ppb range. In a recent development, a parallel affinity sensor array (PASA) system based on chemiluminescence labels (peroxidase/lumonol) and CCD detection was developed (Weller *et al.* 1999) for monitoring pesticides contaminants in water. A detection limit of this system down to 20 ng/L was achieved for terbutylazine. The regeneration of the assay could be performed more than 100 times. A review focusing various immunosensors based on fluorescence detection techniques for pesticide monitoring has been presented by Schobel *et al.* (2000).

1.2.2. Non-labeled Immunosensor

Non-labeled immunosensors are designed so that the formation of an immunocomplex (antibody–antigen complex) on the transducer surface is directly determined by measuring the physical changes (electrical

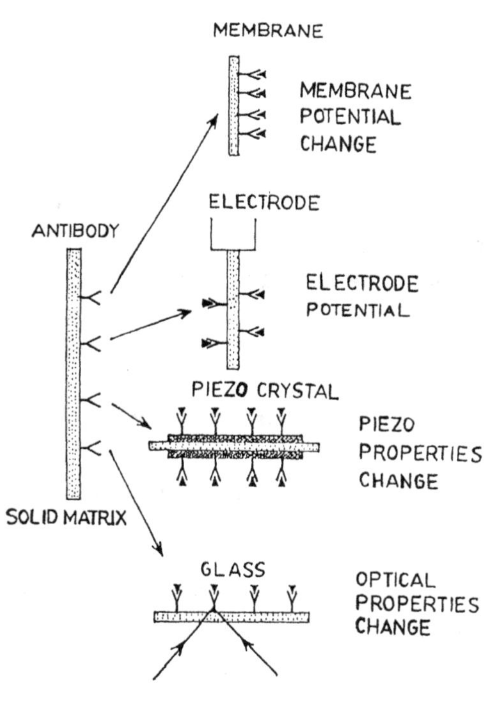

Figure 3. Principle of nonlabeled immunobiosensors.

or optical) induced by the reaction of analytes with the receptor. The advantages of direct immunosensors include rapid detection time and minimal reagents involved. Either the antibody or the antigen is immobilized on the transducer carrier surface to form a sensing device. The solid matrix is sensitive, in its surface characteristics, to detect the immunocomplex formation. The antigen or the antibody to be determined is dissolved in a solution and allowed to react with the complementary matrix-bound antibody or antigen to form an immunocomplex. This formation alters the physical properties of the surface (Figure 3), such as electrode potential (Aizawa *et al.* 1979), piezoelectric properties (Roederer and Bastiaans 1983), or the optical properties (Arnold 1990).

2. TRANSDUCER AND BIOLOGICAL ASPECTS OF IMMUNOSENSORS

2.1. Transducer Aspects

A number of transducers have been demonstrated for immunosensor development. However, three general categories which have widely been explored include optical, piezoelectrical, and electrochemical sensors.

2.1.1. Optical Immunosensors

Optical immunosensors are based on the measurement of a change in the optical characteristics of the transducer surface by the formation of antibody–antigen complexes. Optical sensors can be further classified as direct or indirect. Signal generation in a direct immunosensor depends on the formation of an antibody–antigen complex on the transducer surface. An indirect optical immunosensor may be configured to include various labels such as fluorophores or chromophores to detect the binding. Indirect immunosensing methods produce better signal to noise discrimination, and are therefore quite sensitive for most of applications. However, the sensitivity of the direct methods is limited by nonspecific binding interferents (Robinson 1993). Widely explored optical transducers include fibre-optic, evanescent-wave, and surface-plasmon resonance devices.

2.1.1.1. Fibre-Optic Sensors

Fibre-optic immunosensing technology combines the potential of biochemical recognition by the antibody with the good signal transduction capability of a fibre-optic probe (Arnold 1990). Signal generation can be obtained directly or enhanced by the coupling of a reactant with a label. A fibre-optic strand with an appropriate indicating label at the far end, or tip, of the fibre is illuminated by a suitable light which is introduced into the proximal end. The light travels to the distal tip by total internal reflection. The indicating layer interacts with an analyte of interest and alters the light in proportion to the concentration of the analytes. The main advantage of using fibre optics in immunosensing is their immunity to the effects of electromagnetic radiation, radio frequency interference, harmonic induction and voltage surge, and their potential to perform multiple assays simultaneously, the feasibility of remote sensing and their highly miniaturizable transduction format. Fibre-optic immunosensors have been reported for the estimation of various pesticides in solution. An on-line fibre-optic immunosensor for the detection of atrazine concentration in solution has been demonstrated by Oroszlan *et al.* (1993). Monitoring of triazine derivatives using a fibre-optic-based immunosensor with enhanced sensitivity has been described by Jockers and his colleagues (1993). In this study, selected derivatives were immobilized on the surface of a fibre-optic sensor, and atrazine was determined in a competitive manner using a fluorescein-labeled antibody.

2.1.1.2. Evanescent-Wave Biosensor

When light is propagated through a waveguide (η_1) by multiple total internal reflections an electromagnetic wave called an evanescent wave (EW) is generated in the optically rarer external medium (η_2) with $\eta_1 > \eta_2$. Planar waveguide systems exploit the same optical phenomenon as fibre optics with some minor variation for immunosensing applications (Sutherland *et al.* 1984). The major advantage of this system is minimum possible interference from the bulk media. EW immunosensors have been developed for the analysis of atrazine (Rogers *et al.* 1992).

2.1.1.3. Immunosensors Based on Surface-Plasmon Resonance

Surface plasmon resonance (SPR), a direct optical immunosensing technique, has been successfully incorporated into an immunosensor format for the rapid and non-labeled assay of various biochemical analytes. Substances like proteins, drugs, pesticides or other toxins can be determined directly using either natural antibodies or synthetic receptors with high sensitivity and selectivity. SPR-based immunosensors are capable of real-time monitoring of the antigen–antibody reaction. Immunosensors based on the SPR technique have been developed for the monitoring of various environment pollutants, i.e., pesticides in water and soil (Minunni and Mascini 1993, Chegel *et al.* 1998). This device exploits the properties of the evanescent field and relies on the change in the refractive index of the medium for the generation of signal. In the SPR format, LASER light first enters through the prism at certain angle, and then strikes the glass–metal boundary. The intensity of the reflected light is a function of the refractive index at the glass–metal interface. Any change in the refractive index, caused by the formation of the immunocomplex on metal surface, signifies immunological recognition. A review article by Mullett *et al.* (2000) highlights recent developments in SPR-based immunoassay for low-molecular-weight analytes.

2.1.2. Piezoelectric Crystal Immunosensors

A microgravimetric immunoassay technique using a piezoelectric device as a sensing element is based on the measurement of small changes of mass, resulting from formation of antibody–antigen complex on the crystal surface. The resonant frequency of piezoelectric crystal is mass dependent. Introduction of analyte into the detector cell consisting of modified piezoelectric crystals results in immunocomplex formation, which increases the net mass on the crystal. Consequently, the added mass decreases the vibration frequency of the crystal, which can be quantified to analyte concentration with a detection limit estimated at sub-nanogram levels. The basis of this mass–frequency dependency is attributed to the Sauerbrey equation (Sauerbrey 1959)

$$\Delta F = -2.26 \times 10^{-6} F^2 \Delta m / A,$$

where ΔF is the change of frequency due to the coating (Hz), F is the fundamental frequency of the quartz crystal (MHz), Δm is the mass of deposited coating (g), and A is coated area of crystal (cm^2). The oscillating frequency of the quartz crystal immersed in an aqueous medium is also influenced by the density (ρ) and viscosity (η) of the medium, as described by the equation (Kanazawa and Gordon 1985)

$$\Delta F = -2.26 \times 10^{-6} F^{3/2} (\rho \cdot \eta)^{1/2}.$$

This equation gives a shift in fundamental frequency of the order of 7 kHz for a 10 MHz quartz crystal with only one face in contact with aqueous medium.

Piezoelectric-crystal-based immunosensors for detection of herbicides in water samples have been reported by various groups. A piezoelectric crystal immunosensor for atrazine monitoring in drinking water has been developed (Guilbault *et al.* 1992). Anti-atrazine coated crystal for monitoring atrazine concentrations in solution has been described by Yokoyama and his team (Yokoyama *et al.* 1995). Characterization of monoclonal antibodies to 2,4-dichlorophenoxyacetic acid using a piezoelectric crystal microbalance in solution has been reported (Skladal et al. 1994). However, the use of piezoelectric crystal immunosensors for the detection of small pesticides molecules (M.W. < 1000) is limited because of poor sensitivity of this assay technique for small molecules, and the high level of background response due to nonspecific adsorption. Attempts have been made to use amplification techniques for monitoring small molecules by piezo-crystal-based immunodetection systems (Ebersole and Ward 1988). A direct piezoelectric-crystal-based immunosensor for determination of the herbicide atrazine in solution was developed (Steegborn and Skladal 1997). This immunosensor was used to determine the rate constants of antibody–antigen interaction on modified piezoelectric crystals. The advantages of this technique include moderate cost, usability with flow injection systems, and the potential for detecting large molecules without the requirement for labeling.

2.1.3. Electrochemical Immunosensors

Formation of an antibody–antigen complex on electrochemical transducers alters the ion concentration or electron density on the electrode surface, which in turn can be measured by these electrodes. Electrochemical transducers are broadly classified as amperometric, potentiometric, conductimetric or capacitative transducers, which measure changes in current, potential (voltage), conductance, and capacitance, respectively. Various groups have reported the applications of electrochemical transducers for the detection of different pesticide molecules. A rapid assay based on an immunoenzyme electrode and peroxide conjugates was developed for 2,4-dichlorophenoxyacetic acid (2,4-D) and 2,4,5-trichlorophenoxyacetic acid (2,4,5-T) (Dzantiev *et al.* 1996). The assay monitors the competitive binding of free pesticide and pesticide–peroxide conjugate with anti-pesticide antibody immobilized on a graphite electrode by measurement of peroxide activity in the immune complex on the electrode surface. Detection limits for 2,4-D and 2,4,5-T were about 50 ng/ml with no interference with serum protein present in solution. A simple electrochemical immunosensor-based assay for the field-based quantification of the herbicide 2,4-D in soil extracts was presented by Kroger and colleagues (Kroger *et al.* 1998). The sensor utilizes a competitive immunoassay format incorporating an immobilized antigen complex at the surface of a disposable screen-printed working electrode element. The extent of glucose oxidase-labeled antibody binding to the antigen-electrode is determined amperometrically and is related to sample analyte concentration. A disposable immunosensor with liposome enhancement and amperometric detection technique for monitoring herbicide triazine in real samples was presented (Baumner and Schmid 1998). Hapten-tagged liposome entrapping ascorbic acid as a marker molecule was chosen for the generation and amplification of the signal. Response time of the developed assay was 1–3 min, with sensitivity of measurement in tap water below 1 μg/L of atrazine. In a different approach, an electrochemical biosensor based on acetylcholine (ACh) receptor was developed (Eldefrawi *et al.* 1988). The ACh receptor was fixed to the gate of an ion-selective field effect transistor (ISFET). Binding of ACh with receptor results in a potential change, which is detected with the ISFET.

2.2. Biological Aspects

Pesticides, organic compounds of molecular weight less than 1000, are usually non-immunogenic, and hence do not elicit an immune response unless coupled with some macromolecules such as proteins. It is therefore necessary to modify these small substances (hapten) for coupling with macromolecules (carrier) so as to make a stable carrier–hapten complex. The developed protein–hapten conjugate, which is immunogenic in nature, can be used to generate antibody against pesticides. The work necessary for antibody generation for pesticide molecules is summarized below.

2.2.1. Antibody Development

One of the major challenges in constructing a practical immunobiosensor for pesticides monitoring is the production of a specific antibody against these smaller molecular weight toxins. Most of these pesticide molecules are small (Mr <1000), unable to produce immunoresponse, and require conjugation with carrier proteins to make them immunogenic.

Antibodies are Y-shaped protein molecules, which fall under the general category of immunoglobulins. These are produced and secreted by plasma cells in response to the specific antigen, and are able to recognize and bind antigens with high specificity. Antibody molecules include heavy- and light-chain units, which contain a hypervariable region responsible for the specific recognition of an antigen.

To generate antibodies against these low-molecular-weight substances, the molecules are first modified to link with the carrier proteins to make them immunogenic. The well-characterized protein–hapten conjugates are then used to immunize animals.

2.2.1.1. Hapten Design

Design of hapten for linking with carrier proteins is the most important aspect of antibody generation against pesticides molecules. These molecules are synthesized in such a way that they mimic the structure of the compound and contain a reactive group that can form a covalent linkage with the carrier proteins. Minimal structure alteration to the hapten ensures that the unmodified hapten will be recognized by the antibody in the assay system. The hapten should also largely preserve the chemical structure and spatial conformation of the target compound. Hapten design is critical to ensure the successful generation of antibodies having the desired selectivity. Careful hapten selection may be responsible for either class-specific or compound-specific antibody for the immunosensor system. Specific strategies for hapten design have been discussed extensively in the literature (Harrison *et al.* 1988, Goodrow *et al.* 1990).

2.2.1.2. Conjugation of Hapten to Carrier Proteins

The characteristics of the hapten–protein conjugate influences the generation of specific antibodies against the target molecules. It is therefore essential that the hapten be accessible to generate an immune response. The functional group of the hapten governs the selection of the conjugation method to be developed. Common procedures use amine, carboxylic acid, hydroxyl, or sulfhydryl groups on the hapten and the protein. The most frequently used carrier proteins for conjugation are: bovine serum albumin (BSA), ovalbumin (OVA), conalbumin (CONA), thyroglobulin (TG), immunoglobulin (Ig), fibrinogen, or keyhole limpet hemocyanin (KLH). Hapten–protein conjugates of serum albumin are, in general, more soluble than conjugates of γ-globulin or of ovalbumin. Steroid–protein conjugates of bovine, rabbit, and human serum albumin were found to be soluble above pH 5.5. However, similar conjugates made of γ-globulin and egg albumin precipitated out of solution during preparation.

It is still not clear to what extent the carrier influences the anti-hapten response. Regardless of the protein carrier used, the same functional groups are available for attachment to the hapten, i.e., the carboxyl groups of the C terminal and of the aspartic and glutamic acid residues, the amino groups of the N terminal and the lysine residues, the imidazo and phenolic functions of the histidine and throsine residues, and the sulfhydryl group of cysteine residues. These molecules have been used to make the hapten–protein conjugates. In general, KLH protein, very large in size and usually highly immunogenic, may enhance the antihapten response as a result of the general stimulation of the immune system. For a good immune response, a high hapten density is desirable. Verification of the coupling reaction and determination of the hapten density can be accomplished mainly by evaluating free lysine residues before and after conjugation with spectrophotometry (Habeeb 1966), by matrix-assisted LASER desorption mass spectrometry to determine the mass change before and after conjugation (Hillenkamp 1992), or by radiolabelled haptens (Rinder and Fleeker 1981).

2.2.1.3. Immunization Protocol

Commonly used species for the production of polyclonal antisera are rabbit, guinea pig, sheep, goat, and horse. However, rabbits are the first choice for most purposes unless very large amounts of serum are needed because they are cheap, easy to care for, robust in the face of intensive immunization, and easy to bleed. However, for monoclonal antibody generation, the animal of choice is mice or rats, because of the availability of myeloma cell lines as fusion partners. Production of polyclonal antibodies is relatively simple. However, monoclonal antibodies offer the advantage of containing a defined antibody type, which is produced by established cell lines, allowing unlimited quantities of antibodies as long as the cell line is stable. The choice between monoclonal or polyclonal antibodies depends upon various aspects of the assay like: cost of development, relative specificity and sensitivity of the antibodies, ease of screening and selection of antibodies, etc.

Different immunization procedures are followed for the generation of antibodies against target molecules depending largely on animal variability, type of antigen, and purity of immunogens used. However, a normal protocol for generating polyclonal Abs involves injecting immunogen mixed with the adjuvant (complete Freund to enhance the immune response) intradermally at different spots. The dose rate varies with animal type (a range of 50–1000 μg generally covers all needs). The subsequent booster doses of immunogen with the incomplete adjuvant are then administered at a 3 to 4 weeks time interval. After each booster dose serum samples are collected to monitor the antibody titer. The purification of antibody from the collected serum is done using standard affinity purification techniques, i.e., saturated ammonium sulfate cut and protein-A column chromatography techniques. Hurn and Chantler (1980) have provided a detailed description of the production of antibodies against reagents with major emphasis on immunization protocols.

2.2.2. Immunosensor-based Assay Development

An immunosensor usually contains a specific antibody coated on transducer carrier surface, which binds to the analyte, producing a detectable signal. Depending upon the type of analytes being detected, and also the transducer employed, one of the following assay formats is used for the immunosensor-based analysis:
(a) direct immunoassay;
(b) competitive immunoassay;
(c) sandwich immunoassay;
(d) displacement immunoassay.

With the direct assay technique the molecular interaction can be monitored without the use of labels (Suri *et al.* 1995). The analyte to be detected has innate detectable characteristics by virtue of its mass or optical parameters of the bio-interface of the transducer surface. The direct interactions of small molecules do not produce significant shifts in physical parameters of mass/optical transducer devices, and hence the sensitivity of this type of assay for low-molecular-weight pesticide molecules is insufficient.

Competitive assays monitor the competition between labeled and unlabeled antigen for limited available binding sites of antibodies on the transducer surface. By using a fixed amount of immobilized antibody and labeled antigen, an inhibition curve can be established by introducing different amounts of unlabeled antigen. The concentration of unlabeled antigen in a test sample can be determined by comparison with the inhibition curve (Rabbany *et al.* 1994).

In sandwich assays, the antigen is sandwiched between two different antibodies (preferably with specificity for different epitopes against antigen) on the transducer carrier surface. The first antibody, immobilized on the carrier surface, recognizes the antigen, which further binds to the second antibody at its alternate site. Because of higher sensitivity and low cross-reactivity, this approach is widely employed for detecting low-molecular-weight substances.

Another assay approach, i.e., displacement immunoassay format, immobilized antibody is saturated with labeled antigen. For highest sensitivity, the labeled antigen should be univalent so that it is bound to the carrier by a single epitope. On introducing the unlabeled antigen in a sample, the bound–labeled antigen is displaced from the immobilized antibody, which can be monitored by different transducer systems.

For pesticide analysis by immunosensors, fluorescent labels such as fluorescein isothiocyanate (FITC), fluorescein, or red light emitting dye, Cy5.5, are most widely used. Klotz and colleagues developed an optical immunosensor based on measurement of either fluorescence excitation or emission via the evanescent field of the waveguides, which allow real-time monitoring of the labeled antibodies (Klotz *et al.* 1998). The detection limits of the assay were in the sub-ppb range.

3. CONCLUSIONS

Immunosensor-based analytical techniques provide a simple, inexpensive, and sensitive detection method for monitoring a large number of analytes, including low-molecular-weight pesticide molecules. These techniques are now gaining acceptance, and are competing successfully with other established analytical methods for pesticide determination. However, substantial improvements in this technology are still desired. Generation of specific antibodies against other pesticide molecules, improvement in the electronic transducer system, improved methods to reduce cross-reactivity and background noise, and improvement in the detection limit are some of the parameters which still need to be explored before such techniques become alternatives for established analytical techniques currently being used for the analysis of pesticides in samples. Developments in such areas are likely to result in a higher degree of commerciability in the near future.

REFERENCES

Aizawa, M., S. Kato and S. Suzuki (1977). Immunoresponsive membrane I. Membrane potential change associated with an immunochemical reaction between membrane-bound antigen and free antibody. *J. Membrane Sci.*, **2**, 125–32.

Aizawa, M., S. Suzuki and Y. Nagamura (1979). An enzyme immunosensor. *Anal. Biochem.*, **94**, 22.

Arnold, M.A. (1990). Fiber-optic biosensors. *J. Biotechnol.*, **15**, 219–228.

Baumner, A.J., and R.D. Schmid (1998). Development of a new immunosensor for pesticide detection: a disposable system with liposome-enhancement and amperometric detection. *Biosensors Bioelectron.*, **13**, 519–529.

Chegel, V.I., Y.M. Shirshow, E.V. Piletskaya and S.A. Piletsky (1998). Surface Plasmon Resonance sensor for pesticide detection. *Sensors Actuators B*, **48**, 456–460.

Clark Jr, L.C., and C. Lyons (1962). Electrode system for continuous monitoring in cardiovascular surgery. *Ann. N.Y. Acad. Sci.*, **102**, 29–45.

Dzantiev, B.B., A.V. Zherdev, M.F. Yulaev, R.A. Sitdikov, N.M. Dimitrieva and I.Y. Moreva (1996). Electrochemical immunosensors for determination of pesticides 2,4-D and 2,4-T. *Biosensors Bioelectron.*, **11**, 179–185.

Ebersole, R.C., and M.D. Ward (1988). Amplified mass immunosorbent assay with a quartz crystal microbalance. *J. Am. Chem. Soc.*, **110**, 8623–8628.

Eldefrawi, M.E., S.N. Sherby, A.G. Andreou, N.A. Mansour, Z. Annau, N.A. Blum and J.J. Valdes (1988). ACh receptor based biosensor. *Anal. Lett.*, **21**, 1665–1680.

Goodrow, M.H., R.O. Harrison and B.D. Hammock (1990). Hapten synthesis, antibody development, and competitive inhibition enzyme immunoassay for s-triazine herbicides. *J. Agric. Food Chem.*, **38**, 990–996.

Guilbault, G.G., B. Hock and R. Schmid (1992). A piezoelectric immunobiosensor for atrazine in drinking water. *Biosensors Bioelectron.*, **7**, 411–419.

Habeeb, A.F.S.A. (1966). Determination of free amino groups in proteins by trinitrobenzenesulfonic acid. *Anal. Biochem.*, **14**, 328–336.

Harrison, R.O., M.H. Goodrow, S.J. Gee and B.D. Hammocks (1988). Immunochemical methods for pesticide residues analysis. In *Biotechnology in Crop Protection*. P.A. Hedin, J.J. Menn and R.M. Hollingsworth (Eds.), *ACS Symp. Ser.*, **379**, ch. 24.

Hartley, I.C., and J.P. Hart (1994). Amperometric measurement of organophosphorous pesticides using a screen printed disposable sensor and a biosensor based on cobalt phthalocyanine. *Anal. Proc. Including Anal. Commun.*, **31**, 333–337.

Hillenkamp, F. (1992). Determination of the hapten density of immunoconjugates by MALDI mass spectrometry. *Anal. Lett.*, **25**, 1983–1997.

Hudson, L., and F.C. Hay (1989). In *Practical Immunology*, 3rd edition, Elaine and Frances (Eds.), Blackwell Scientific, Oxford, pp. 207–263.

Hurn, B.A.L., and S.M. Chantler (1980). Production of reagent antibodies. *Methods Enzymol.*, **70**, 104–142.

Imato, T., and N. Ishibashi (1995). Potentiometric butyrylcholine sensor for organophosphorous pesticides. *Biosensors Bioelectron.*, **10**, 435–441.

Jockers, R., F.F. Bier and R.D. Schmid (1993). Enhancement of immunoassay sensitivity by molecular modification of competitors. *J. Immunol. Methods*, **163**, 161–167.

Kabat, E.A. (1976). *Structural Concepts in Immunology and Immunochemistry*, 2nd edition. Holt, Rinehart and Winston, New York, pp. 24.

Kanazawa, K.K., and J.G. Gordon (1985). The oscillation frequency of a quartz crystal resonator in contact with a liquid. *Anal. Chim. Acta*, **175**, 99–105.

Klotz, A., A. Brecht, C. Barzen, G. Gauglitz, R.D. Harris, G.R. Quigley, J.S. Wilkinson and R.A. Abuknesha (1998). Immunofluorescence sensor for water analysis. *Sensors Actuators B*, **51**, 181–187.

Kroger, S., S.J. Setford and A.P. Turner (1998). Immunosensor for 2,4-dichlorophenoxyacetic acid in aqueous/organic solvent soil extracts. *Anal. Chem.*, **70**, 5047–53.

Marty, J.L., K. Sode and I. Karube (1992). Biosensor for detection of organophosphorous and carbamate insecticides. *Electroanalysis*, **4**, 249–252.

Minunni, M., and M. Mascini (1993). Detection of pesticides in drinking water using real-time biospecific interaction analysis (BIA). *Anal. Lett.*, **26**, 1441–1460.

Mionetto, N., J.L. Marty and I. Karube (1994). Acetylcholinesterase in organic solvents for the detection of pesticides: biosensor application. *Biosensors Bioelectron.*, **9**, 463–470.

Mullett, W.M., E.P. Lai and J.M. Yeung (2000). Surface plasmon resonance-based immunoassays. *Methods*, **22**, 77–91.

Muramatsu, H., J.M. Dicks, E. Tamia and K. Isao (1987). Piezoelectric crystal biosensor modified with protein A for determination of immunoglobulin. *Anal. Chem.*, **59,** 2760–2763.

Oroszlan, P., G.L. Duveneck, M. Ehrat and H.M. Widmer (1993). Fiber-optic atrazine immunosensor. *Sensors Actuators B*, **11**, 301–305.

Rabbany, S.Y., B.L. Donner and F.S. Ligler (1994). Optical immunosensors. *Crit. Rev. Biomed. Eng.*, **22**, 307–346.

Rinder, D.F., and J.R. Fleeker (1981). A radioimmunoassay to screen for 2,4-D and 2,4-T in surface water. *Bull. Environ. Contam. Toxicol.*, **26**. 375–380.

Robinson, G.A. (1993). Optical immunosensors. In *Methods of Immunological Analysis*, Vol. 1: Fundamentals. R.F. Masseyeff, W.H. Albert and N.A. Staines (Eds.), Wiley, Canada, pp. 371–388.

Roederer, E., and G. Bastiaans (1983). Microgravimetric immunoassay with piezoelectric crystals. *Anal. Chem.*, **55**, 2333–2336.

Rogers, K.R., N.A. Anis, J.J. Valdes and M.E. Eldefrawi (1992). *ACS Symp. Ser.*, **511,** 165.

Sauerbrey, G. (1959). The use of oscillators for weighting thin layers and for microweighting. *Z. Phys.*, **155**, 206–212.

Schobel, U., C. Barzen and G. Gauglitz (2000). Immunoanalytical techniques for pesticide monitoring based on fluorescence detection. *Fresenius J. Anal. Chem.*, **366**, 646–658.

Skladal, P., M. Minunni, M. Mascini, V. Kolar and M. Franek (1994). Characterization of monoclonal antibodies to 2,4-dichlorophenoxyacetic acid using a piezoelectric quartz crystal microbalance in solution. *J. Immunol. Methods*, **176**, 117–125.

Steegborn, C., and P. Skladal (1997). Construction and characterization of the direct piezoelectric immunosensor for atrazine operating in solution. *Biosensors Bioelectron.*, **12**, 19–27.

Suri, C.R., P.K. Jain and G.C. Mishra (1995). Development of PZ crystal based microgravimetric immunoassay for determination of insulin concentration. *J. Biotechnol.*, **39**, 27–34.

Sutherland, R.M., C. Dahne and J.F. Place (1984). Preliminary results obtained with a no label homogeneous, optical immunoassay for human immunoglobulin G. *Anal. Lett.*, **17**, 43–53.

Tran-Minh, C., P.C. Pandey and S. Kumaran (1990). Studies of acetylcholine sensor and its analytical application based on inhibition of cholinesterase. *Biosensors Bioelectron.*, **5**, 461–471.

Weller, M.G., J.S. Andreas, M. Winklmail and R. Niessner (1999). Highly parallel affinity sensor for the detection of environmental contaminants in water. *Anal. Chim. Acta*, **393**, 29–41.

Yokoyama, K., K.I. Ikebukuro, E. Tamiya, I. Karube, N. Ichiki and Y. Arikawa (1995). Highly sensitive quartz crystal immunosensors for multisample detection of herbicides. *Anal. Chim. Acta*, **304**, 139–145.

INDEX

Advances in Biosensors
Volume 5, pages 179–186.

www.ingramcontent.com/pod-product-compliance
Lightning Source LLC
LaVergne TN
LVHW011803200926
848484LV00003B/768

9780444513373